IB CHEMISTRY

Internal Assessment

IB CHEMISTRY

Internal Assessment

For the International Baccalaureate Diploma

Zouev Elite Publishing

Published 2021

Printed by Zouev Elite Publishing

ISBN 978-1-9996115-2-1, paperback.

TABLE OF CONTENTS

PART I

THE CHEMISTRY IA GUIDE

1. GENERAL INTRODUCTION

The Internal Assessment (IA) is a significant component of the overall assessments, composing 20% of your overall grade. In terms of points, it is worth $0.2 \times 7 = 1.2$ points![1] Comparing this to the extended essay, which together with TOK is worth 3 points: $0.5 \times 3 = 1.5$. Yet, many students spend significantly more time on their EEs than their IAs. According to the subject guides, the investigation itself should take approximately 10 hours, which includes time in discussion with your teacher/supervisor. Following this you should make a writeup of about 6 – 12 pages long. Note that the examiner is not required to read anything beyond 12 pages, so anything past that point is potentially ignored. The 12 pages do not include appendices and bibliography.

The purpose of the IA is to examine your ability to take concepts you have studied theoretically, and apply them to a research question of your choice. You are aiming to show your ability to think critically in terms of planning an experiment, as well as reflecting on the analysis and implications of your results. It is also meant to serve as an application and extension of the laboratory skills you acquired in the labs experiments you have performed as part of the curriculum teaching.

Examiners are looking for demonstration, to various degrees, of the ten qualities identified in the IB learner profile. These qualities are assessed in the more concrete and specific requirements of the IA's assessment criteria, so these should be borne in mind at all times when planning and writing the IA.

Barring unusual circumstances, such as the May 2020 exam season[2], IAs are marked **internally** *i.e.,* by your teacher, then **moderated** by an external examiner. This has the following implications:

1. **Internal assessment**: Your teacher is the one with the most influence on the grade you are assigned. Ensure that you are writing in line with your teacher's expectations in terms of formatting, notation, page limits, citations, and any other peculiarities. For example, in the procedures/methods section, some teachers prefer a numbered list of steps for easy reading, while others favour prose-style writing in line with scientific journals. In any case where there is a conflict between the advice in this book and your teacher's requirements, follow your teacher's requirements!

2. **External moderation**: This is where a random sample of IAs from each school is sent to an external IB examiner, for independent marking. If the examiner's marking differs from the teacher's by more than two points, the entire cohort is *moderated, i.e.,* the grades are adjusted up or down. This means that while you are given a grade by your teacher, there is a chance that this will be adjusted by the IB Organisation (IBO). Note that the adjustment involves an algorithm by the IBO which is not published, and each student's grade may not be moved up or down by the same amount.

---

[1] For May 2021 sessions, the assessment weightings have been revised to take into account the removal of the Paper 3 examination. However, the IA weighting remains as 20% of the final grade, with the proportion of Paper 3 going to Papers 1 and 2.
[2] In this season the exams were cancelled and all IAs were marked externally.

In the rest of this document, we will first lay out a recommended structure for the report. This structure is a standard format for scientific writing - most reports, from master's theses to scientific journals, will follow this in some form. However, for many students, this may be the first time they are required to produce a detailed report, so this structure will be of some use as a starting framework. This framework will also include more detailed discussions of what should go into each section, and how to address each of the points required.

Next, we will examine the Assessment Criteria for the IA in detail, which provides a rubric for teachers and examiners to grade the internal assessment. This is freely available in the syllabus of the subject, but quite often students are not directed to this (or they syllabus!). There will also be notes on how to meet the criteria, which include which sections of the report structure serve to address them. Checking your IA against the criteria in the rubric as you write will ensure that you will receive a good result for the internal assessments.

Finally, we will close out with some miscellaneous tips and things to bear in mind, which did not fall under any of the previous sections.

2. IA REPORT STRUCTURE

Laid out here is a recommended structure for your IA report, which is similar to the structure of most forms of scientific writing. Following this structure will ensure that you have dedicated sections to address each of the points in the grading criteria, as well as a sensible flow for the report, which is vital for picking up the marks in the "Communication" criteria. The structure is followed by a detailed description on the purpose of each section and what you should aim to cover. As always, if you have a prescribed structure from your teacher, place your content into the relevant equivalent sections of their structure.

Sample Structure

Introduction

- Motivation

- Research question

- Background theory

- Hypothesis

Materials and Methods

- Variables

  o Controlled

  o Independent

  o Dependent

- Apparatus/Materials
- Diagram of apparatus
- Methodology
- Risk assessment
  - Safety hazards
  - Environmental hazards
  - Ethical concerns

Data Collection and Analysis

- Raw data:
  - Quantitative (tables)
  - Qualitative (notes/observations)
- Example calculation for analysis
- Example uncertainty calculation
- Processed data
  - Tables
  - Graphs and other charts
  - Discussion of results with each chart

Conclusion and evaluation

- Conclusion
  - Review of results and comparison to hypothesis
  - Answer the research question as much as possible
  - State if results support the hypothesis.
- Evaluation
  - Comparison of result with literature
  - If conclusion disagrees with hypothesis or literature, why?
  - Limitations of experiment
  - Sources of error
  - Possible improvements

        o   Future investigations/extensions

Bibliography

Appendix

3. PURPOSE OF SECTIONS

Introduction

The introduction serves to lay the background and the context of the investigation, both in terms of the scientific understanding, as well as the reasons that you are investigating the particular experiment.

The <u>motivation</u> subsection states the personal reason you chose this topic. Perhaps it is related to one of your hobbies, or it drew your interest when you learnt about it through your course or investigations. Keep this section short: no longer than one paragraph! You do not need to tell a long story. The personal interest is only one small part of the personal engagement criteria. For example, you may say something like,

"I regularly take vitamin supplements in the form of effervescent tablets. I was interested in how the mixing of the tablet with water could produce gas so quickly, and so wished to investigate the rates of reaction between the ingredients causing the release of CO_2"

would suffice.

Dedicate a subsection to your <u>research question</u>: the explicit statement of the research question is a concrete criteria which is easily achieved. Make sure it is expressed as a question, such as,

"How does raising the concentration of Cl in the reaction with $AgNO_3$ affect the amount and rate of precipitate production?".

For more information on how to choose your research question, see the "Choosing Your Research Question" in the final tips section.

<u>Background theory</u> is where you will lay out the theoretical information required for analysing and understanding the experiment. This should be quite a lengthy section, typically spanning about two pages, as you will need to include a summary of the key concepts of the theory, and how they apply to the specifics of the experiment. For example, if you were conducting a kinetics experiment, you would describe the principles of reaction rates and the rate expression, the particular chemical reaction you are investigating, and how its molecularity and mechanism may be a predictor of the rate expression. You would likely wish to go into particulars of how to calculate the order of reaction and activation energy from the experimental data which you intend to collect, with the relevant equations and manipulations. In this part, use general formulas and equations, rather than the data you have generated. Putting the data into the calculations should be left for the data analysis section.

Lay out your <u>hypothesis</u> clearly, in terms of what you expect to find, together with the reasoning. This may be based on a particular formula from literature, derived from formulas from different sources, or based off everyday observation and intuition. Remember to state the expected results and your expected general answer to the research question, *e.g.*,

"From the proposed mechanism of the reaction, I would expect the reaction to be first order with respect to NaOH. Therefore, I would expect a linear increase in the rate of reaction as the concentration of NaOH increases until I reach the point where the surface area of the carbonate becomes the limiting factor".

Materials and Methods

Lay out your <u>variables</u> and classify them as either **controlled** (what you will keep constant throughout the experiment), **independent** (what you will intentionally vary), and **dependent** (what you measure to change due to varying the independent). It may be useful to use a table such as the following:

Variable	Type	Reason	How to control
Temperature	Independent	The rate constant is dependent on temperature. Measuring the rate of reaction at different temperatures allows the calculation of activation energy	Use different temperature settings on the water bath and ensure the equipment has reached the thermal equilibrium before beginning the reaction.
Surface Area of marble chips	Controlled	The surface area of the solid reactant affects the rate of reaction.	Try to use marble chips of approximately the same size.

Note that in the example of the controlled variable, the surface area of the marble chips, we are unable to fully control the surface area since there will be natural variability in the materials. This is worth noting as a source of error/uncertainty, and doing so will demonstrate both reflection on experimental limitations, and personal engagement.

<u>Apparatus/Materials</u> is a somewhat optional section; a well-detailed procedure would include all the information here. However, it can be useful to state the concentrations of solutions and types of chemicals involved in the reaction. If you have a series of solutions of different concentration, you can also lay out how these different concentrations were prepared in this section rather than in methodology if you wish.

A <u>diagram</u> of the apparatus and experimental setup is very useful both in helping the reader to visualise the experiment, as well as save on text in the procedure section. You can either draw your own diagram of the experiment, or as is much easier these days, take a photo of the experiment as you are performing it, and annotate it appropriately.

Lay out the process step-by-step in the <u>methodology</u>. You can either use a numbered list of steps (preferred about 90% of the time), or describe the procedure in prose. Remember to detail which trials are repeated, in order to ensure that you have collected sufficient data (*"sufficient relevant quantitative and qualitative raw data"*). Usually, if you are plotting a trend you would need at least five points, and each of those points would need three repeats, so you'd be looking at about 15 runs in total. **Plan your timepoints and concentrations** so that you can perform these 15 runs in the lab time available to you. It may be useful to do a pilot trial to see how quickly a reaction runs, or how much volume is required to titrate before finalising the details and concentrations of the experiment.

Use a risk assessment to address any **reasonable** potential dangers, either in terms of safety, or environmental or ethical concerns. Describe the source and its hazards, and how to mitigate them. The following table layout may be useful:

Description	Hazard	Preventative measures
Hydrochloric Acid (1M)	Corrosive- can cause chemical burns if in contact with skin or eyes	Wear lab coat, gloves, and goggles when handling. Handle in fume hood as much as possible
Broken glassware	May cause cuts/injuries when handled	In case of broken glassware, handle with care and inform lab technician or demonstrator immediately
Chlorinated solvents	Toxic; Environmental impact if allowed to enter the environment: harmful to aquatic organisms and suspected carcinogens	Handle with lab coat, gloves, and goggles. Separate chlorinated solvents and dispose of according to local regulations.

Generally, chemicals will have potential health and environmental impacts, but negligible ethical impact (unless there are ethical issues with the source of the chemical!). Show that you have considered and dismissed the ethical impact by stating that there are no significant ethical implications to your experiment. Of course, if there are no significant safety hazards in the chemicals or experiments used, you can make a similar statement. The key point is to show that you have given these issues due consideration.

Data Collection and Analysis

In the raw data section, lay out the **unprocessed data** you collect. This would be usually in the form of a table such as:

	Absorbance/% +/- 0.1%								
Conc. HCL	0.10 +/- 0.01 M			0.20 +/- 0.01 M			0.30 +/- 0.01 M		
Time +/- 0.1s	Trial 1	Trial 2	Trial 3	Trial 1	Trial 2	Trial 3	Trial 1	Trial 2	Trial 3
10									

20								

Note the inclusion of the absolute uncertainty in the headers of the table. If the uncertainty of each reading is different, you will need additional columns to place the specific uncertainty next to each datum. If you have a very large amount of data due to multiple trials and runs, you may choose to put this table in the appendix, and place a summary table in the raw data section, which could include just the averaged values. If you do, state that the raw data has been placed in the appendix.

In this section, also include qualitative results, describing any observations, such as

"The colour of the solution was observed to change from blue to orange but became colourless after 10 minutes", *"Trial 1 did not show significant reaction"*, or *"the osmotic tubing ruptured in trial 5 so the results could not be obtained"*.

Qualitative observations will give support and rationale for discarding data points which may be outliers, or just missing.

Include an <u>example calculation for analysis</u>: this is simply an example calculation you made for one of the trials, written out step-by-step in detail. For example:

$$\text{average volume of titre} = \frac{\text{trial 1 + trial 2 + trial 3}}{3}$$

$$= \frac{0.24 + 0.25 + 0.21}{3} = 0.233 \text{ml}$$

$$\text{moles HCL} = \text{conc. HCL} \times \text{vol. HCL} = 0.1 \times 0.233 = 0.0233 \text{ mol}$$

$$\text{conc NaOH} = \frac{\text{mol NaOH}}{\text{vol NaOH}} = \frac{0.0233}{0.025} = 0.932 \text{ mol dm}^{-3}$$

All other calculations can then be presented in the analysis table without working in the <u>processed data</u> tables. Use your spreadsheet software, such as Excel, to perform the calculations on the data easily, so you do not have to spend the time to do so manually.

You will also need to do an <u>example uncertainty calculation</u>. This is presented in the same way as the example calculation. The uncertainties should be calculated according to the principles in Topic 11: Measurement, Data Processing, and Analysis. An example would be:

$$\text{abs. uncertainty of ave vol} = \frac{\sum \text{abs. uncertainty of trials}}{3} = \frac{0.01 + 0.01 + 0.01}{3} = 0.01$$

variability of trials greater than calculated abs uncertainty: largest variability taken as uncertainty:

$$0.233 - 0.21 = 0.023$$

$$\% \text{ uncertainty of ave vol} = \frac{\text{abs unc.}}{\text{value}} \times 100 = \frac{0.023}{0.233} \times 100 = 9.8\%$$

$$\% \text{ uncertainty of moles. HCL} = \% \text{ unc. conc HCL} + \% \text{ unc. vol HCL} = 1\% + 9.8\% = 10.8\%$$

In the processed data section include the calculated data in the form of a table. The table of analysed data can similarly be a summarised one, rather than one including all your intermediate calculations, as you have already laid out your calculations in the example calculation section. Also include in the processed data section any graphs or charts that you draw from these data. See the next section on what types of charts to plot, and how to analyse them.

Linearization of Data

Generally speaking, you will be trying to plot **linear** graphs. There are advantages to this:

- It is easy to see if the hypothesis is correct if the data points on the graph show a linear relationship (or not!).

- It is the easiest type of graph to determine the uncertainty of fit from, using maximum and minimum gradients.

- The quality of the fit given by Pearson's Product Moment Correlation Coefficient (PPMCC) can be used to assess the quality of the regression and fit.

Since relationships between dependent and independent variables are often not linear, you may need to manipulate the equations relating the variables into $y = mx + c$ form. A typical example is the Arrhenius equation:

$$k = Ae^{-\frac{E_a}{RT}}$$

You may be trying to relate k (dependent variable) to T (independent variable) and plot a graph. However, k (x-axis) and T (y-axis) are not in the form of a linear graph. In this case, we perform the following manipulations:

$$\ln k = \ln A + \ln \left(e^{-\frac{E_a}{RT}}\right)$$

$$\ln k = \ln A - \frac{E_a}{RT}$$

$$\ln k = -\frac{E_a}{R} \times \frac{1}{T} + \ln A$$

Comparing the equation to $y = mx + c$, we can see that they now have a similar shape. If we let the dependent axis $y = \ln k$, and the independent axis $x = \frac{1}{T}$, we get gradient $m = -\frac{E_a}{R}$ and y-intercept $c = \ln A$. The key to this is that you do not need to plot the dependent and independent variables directly on the axis, and plotting some $f(x)$ and $g(y)$ on their respective axes will yield a linear graph that is amenable to regression on your graph plotting software. Recall from your maths course that Pearson's Product-Moment Correlation Coefficient, r, is only meaningful for linear fits.

Uncertainties in Linear Fits

Linear fits also allows for determination of maximum and minimum gradients if you are using the gradient or y-intercept to extract a value. These should be plotted using the error boundaries of the extreme ends of the data.

Here is an example graph of $y = 2x$ where x and y both have an uncertainty of ± 0.5.

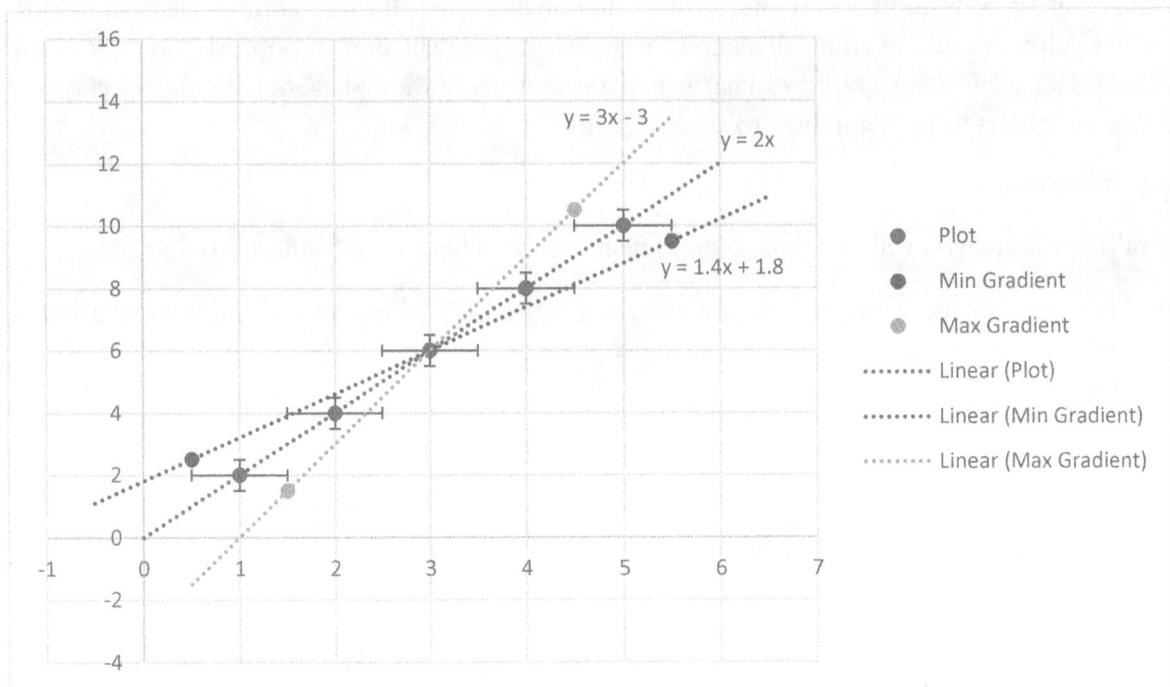

The blue trendline is the fit of the data; the error bars have been included for these data points. For the minimum gradient, shown in orange, plot the upper left corner of lowest data point, and the bottom right corner of the highest data point. Conversely, for the maximum gradient, plot the bottom right corner of the lowest data point, and the upper right corner of the highest data point. Your working spreadsheet would look something like:

Plot				Min Gradient	x	y		Max Gradient	x	y
x +/- 0.5	y +/- 0.5									
1	2			Lowest point	0.5	2.5		Lowest point	1.5	1.5
2	4			Highest Point	5.5	9.5		Highest point	4.5	10.5
3	6									
4	8									
5	10									

Note that we have calculated the values of the corners of the highest and lowest points, and plotted them on the same axes. Fitting the maximum and minimum gradients with a trendline also gives us directly their gradients and their y-intercepts. From this graph, the gradient in that case would be given by $m = 2 \pm 1$, where 1 is the greater difference between the fitted gradient and the min or max gradient.

Once you have plotted the charts, give a short description of what they are showing:

- Is there an increasing or decreasing trend?

- Is the relationship linear, polynomial, exponential, or unclear?

- Is the data clustered together? Close to the trendline?

Discuss if the results are in line with what you were expecting, and their implications. For example, if you are measuring heat capacities of different materials, and you plot temperature (y axis) against time (x axis) state that the steeper gradient has the lower heat capacity since $q = mc\Delta T$ and q is proportional to time if you are supplying a constant power.

Conclusion and Evaluation

In the <u>conclusion</u>, draw together a summary of your data and findings, and state in their context if your hypothesis is supported. Provide an answer to the research question. To reiterate, *your conclusion does not have to support your hypothesis, just be consistent with your findings*. The conclusion in itself is quite short, about the length of a significant paragraph (as opposed to a page).

The <u>evaluation</u> is where you examine any differences between the findings and the hypothesis, and think of reasons why either the hypothesis was mistaken, or if there were confounding factors in the experiment that were not accounted for. A disagreement between the results and the hypothesis gives you more material to work with here; if everything worked according to plan this part of the evaluation would be quite sparse.

In this section, also compare your result to literature values, and if your results are consistent with the scientific consensus and understanding. State if they are in agreement to within uncertainty, or if your results differ significantly. For example, an experimental dissociation constant of water at 298 K of

$$K_w = 9.21 \times 10^{-15} \pm 9\%$$

is in agreement with the literature value of $K_w = 1.023 \times 10^{-14}$, as the literature value falls between the maximum and minimum uncertainty of $K_{w\ min} = 8.3811 \times 10^{-15}$ and $K_{w\ max} = 1.0389 \times 10^{-14}$. A result of $K_w = 9.21 \times 10^{-15} \pm 8\%$ would **not**, as the max value is 9.9468×10^{-15}, and the literature value does not fall within the uncertainty range. As before, disagreement with the literature value provides opportunity for displaying critical thinking and reflection. Consider possible <u>sources of error</u> that may have led to discrepancies, or mistakes in mental models or arguments which you can correct here. From the previous example, perhaps the room temperature was not 25°C as assumed, but higher, leading to a lower actual K_w.

These considerations will lead naturally to discussions of the <u>limitations</u> of the experiment. Discuss how these restrict the results that can be drawn from the investigation. Are they errors systematic (affects accuracy) or random (affects precision)? Did they have a material impact on the results? Propose improvements that could be made if you had more time/equipment to address each of these sources, as well as experiments that would broaden the scope and/or certainty of the conclusion in <u>future investigations/extensions</u>.

Bibliography

List all references you have used here. Try to have at least five references. It is easier that you'd expect (like getting your 5-a-day). Include your Chemistry textbook; cite it for your background theory. Reference any websites which you used to draw theory from, which may add two or three more citations. Referring to databases or scientific journals which provide literature values or experimental significance will easily bring you to five citations. Of course, more citations is never a bad thing. Remember that Wikipedia is not a reliable source, but the references in the articles provide a good place to start searching for supporting evidence.

Appendix

This is where all the expansive data and detailed calculations/derivations which are not crucial to understanding can be placed. Whether this is extensive or non-existent will depend on your investigations and the type and quantity of data generated. For example, if you used a time-logger in the scale of milliseconds and have hundreds of rows of raw data, put that here in the appendix rather than having it take up pages in the raw data section. Similarly, if you have a lengthy step-by-step derivation or manipulation taking more than perhaps a page, place it here, and state just the result in the main body. Make a note in the body text that these resources can be found in the appendix, for those who are interested.

4. GRADING CRITERIA

Introduction

The grading criteria are available in the subject syllabus, under "Internal Assessment". The grades for each of these are assigned separately, from zero to their maximum, based on the descriptions. Note that maximum points **does not imply perfect performance**, just that they have satisfied the criteria in the rubric to achieve those required points. Unfortunately, many teachers are afraid to give maximum points due to this misconception. Put another way, the grading criteria are the equivalent of the mark scheme for the internal assessment: if you fulfil all the criteria in a band, the marker should award you the points for that band. There *are* a few criteria which are differentiated only by the adjectives they use and their interpretation is subjective, such as "some consideration" vs "full and appropriate consideration". However, the majority of them are objective and concrete, so make sure you get as many of them as possible!

The total number of marks is 24, and are broken down accordingly:

- Personal Engagement (2)

- Exploration (6)

- Analysis (6)

- Evaluation (6)

- Communication (4)

Personal Engagement (out of 2 points)

This is the most oft-misunderstood criteria by students (and sometimes teachers). The definition of personal engagement in the syllabus is "*the extent to which the student engages with the exploration and makes it their own.*" This is evidenced in two ways in the assessment criteria, **both** of which must be fulfilled to get the mark at each level:

1. How the research subject is of interest to them, or is directly related to their everyday life: "*personal significance, interest, or curiosity*"

2. What decisions they have made in terms of the "*design, implementation, or presentation of the investigation*".

The mistake many people make with personal engagement is to think that it is only the first point, and dedicate paragraphs to stories describing why the research question is important to them, while neglecting the second point. This allows them to fulfil only one of the two criteria. In this case, even if you address point 1 very strongly, but neglect point 2, you will still receive 0 marks for personal engagement.

In order to ensure that you address point 2, you must constantly show evidence of your personal decision-making. Examples of such places in the investigation would be:

- **Design**: why you have chosen to investigate one case over another, *e.g.*, investigation of one chemical reaction or process over another.

- **Understanding/interest**: what approximations you have made in the theory or calculations, and why.

- **Implementation**: Why you chose the range of dependent variables.

- **Implementation**: Why you have chosen the number of data points taken, the time between readings, etc.

- **Presentation**: How you present the data; why you have chosen to plot particular variables against each other.

Many of these decisions could be made by factors out of your control, *e.g.*, there are only certain materials available in the lab, the thermometer/temperature gauge only takes readings of this range, or the maths becomes incredibly complex without an approximation. However, you have still made a decision based on these restrictions, and so make sure you explain how you have designed and carried the experiment within your limitations to meet the second criteria.

Exploration (out of 6 points)

For the exploration criteria, you are expected to demonstrate that you understand the theory and concepts behind the experiment, its context and risks, and that the science is of an **appropriate** level. The evidence for these criteria is generally presented in their relevant sections, which will be underlined.

"The topic of the investigation is identified, and a relevant and fully focused research question is clearly described."

This should be described and clearly stated in your motivation and your research question. As this is a concrete criteria, dedicate a section in the introduction/motivations to explicitly state your research question.

"The background information provided for the investigation is entirely appropriate and relevant and enhances the understanding of the context of the investigation."

Demonstrate your understanding of the topic by writing a complete but concise background theory section. Usually this will be about 2 - 3 pages. When considering whether to write about a topic, remember that the report should be accessible to another IB student without the relevant background. Another way of putting this is:

- If it is in the prior learning (if you'd have learnt it before the IBDP) you should not include it but only state the results;

- If it is in the course, write about it briefly;

- If it is beyond the course, write about it in detail.

Much of the background theory can be adapted from reference texts and other sources; be sure to cite these sources when you use them and add them to your bibliography. Also, **cite as you write**! Add to your bibliography as you write, especially for online references: it can be difficult to locate the site you got the reference from again afterwards.

"The methodology of the investigation is highly appropriate to address the research question because it takes into consideration all, or nearly all, of the significant factors that may influence the relevance, reliability and sufficiency of the collected data."

The bulk of this will be demonstrated in the <u>materials and methods</u> section. The design and purpose of steps in the experiment should be made clear (another chance to show personal engagement). Remember to include descriptions on the number of repeats per trial, and account for errors and uncertainties. A separate <u>variables section</u> explicitly stating all the controlled, independent, and dependent variables, and the measures to control them, will also go a long way to satisfying this criteria.

"The report shows evidence of full awareness of the significant safety, ethical or environmental issues that are relevant to the methodology of the investigation."

This criteria can be easily fulfilled with a <u>risk assessment section</u> in the materials and methods section. Consider reasonable hazards such as:

- toxic chemicals used

- broken glassware

- heavy weights

- objects under strain or pressure

- strong magnetic fields

Include measures to mitigate or minimise these hazards, such as working in fume hoods, wearing lab coats/eye protection, and warning people with pacemakers to stay away from the magnetic fields. Also describe processes to take should there be spills and how to minimise environmental impact, such as proper disposal of reagents. Ethical issues, if any, should also be addressed in this risk assessment.

If the experiment does not pose any significant risks in these areas, make this statement so show that this aspect has been considered. **Do not include negligible hazards** just so that you have something to talk about: *e.g.*, if you are working on a simulation or theoretical IA, carpal tunnel syndrome and poor ergonomics are not significant hazards: carrying out the IA does not cause you to have a higher risk than routine computer use. The grading rubric explicitly states that this criteria should only be considered should the risks be "*relevant to the methodology of the investigation*".

Analysis (out of 6 points)

The analysis criteria examine how well you have collected, analysed, interpreted, and drawn conclusions from the experiments you have carried out. This will mainly be addressed in the data collection/analysis and conclusion sections.

"The report includes sufficient relevant quantitative and qualitative raw data that could support a detailed and valid conclusion to the research question."

This should be demonstrated with the tables of <u>raw data</u> that you have collected in the <u>data collection section</u>, which should be accompanied by any <u>qualitative observations</u> you have. For example, you may put down a table of temperature, pH, and absorbance of an aqueous compound as your quantitative data. This would be accompanied by a short paragraph about any observations you made which were not represented in the table, such as "*fizzing was observed when the solid was added to the solution*", and "*the colour change was only observed a few seconds after the reagent was added.*" Ensure that the design of the experiment generates enough data, usually by repeating measurements for three or five cycles, depending on time available.

If you have a large amount of raw data, you may choose to put this in the appendix so that they don't take up many pages of the report. If you do so, make reference to the appendix in the data collection section, and make sure to put in a <u>table of summary data</u>, such as averages, in the main text.

"Appropriate and sufficient data processing is carried out with the accuracy required to enable a conclusion to the research question to be drawn that is fully consistent with the experimental data."

Demonstrate the data processing in the <u>data analysis section</u> with an <u>example calculation</u>, paired with the <u>example uncertainty calculation</u>. The full set of processed data can be presented in table form, followed by <u>graphs of the data</u> if necessary. Use graphs to confirm relationships between dependent and independent variables. Try to manipulate equations to give you linear plots, in order to calculate maximum and minimum gradients for your errors. How to do this is described in the "IA Report Structure" section earlier. A linear plot will allow you to make a <u>conclusion</u>, based on if the data is in line with your hypothesis. Note that nowhere in the criteria does it require the data from the experiment to be the correct literature value, or give a result that is in line with current scientific consensus. **It's fine if the conclusions from the data are not "correct"**, as long as the conclusion is consistent with the data, *i.e.*, you state if the results agree/don't agree with the hypothesis. You will have the opportunity to comment on why there is a disagreement in the <u>evaluation</u> section.

"The report shows evidence of full and appropriate consideration of the impact of measurement uncertainty on the analysis."

The <u>analysis section</u> should include an <u>example error calculation</u>, and each datum should have associated errors, whether it is raw or calculated. Include <u>error bars in your graphs</u>, unless they are too small to be seen, in which case make a note that the errors are too small to be visible. This will be used when comparing the results to your hypothesis or to literature values in the <u>evaluation</u>. For example, if your experimental value for g is 9.79 ± 0.05, and the literature value is 9.81, then you will be able to make a statement along the lines of: "the experimental value is in agreement with the literature value within the error of the experiment" since 9.81 falls within the range of the error. If you don't have the error values, you will not be able to make this comparison. The size of the percentage error also allows you to evaluate the precision of your experiment, to be discussed in light of the evaluation criteria.

"The processed data is correctly interpreted so that a completely valid and detailed conclusion to the research question can be deduced."

In the data analysis section, once you have the processed data, take some time to describe what the data is showing you. This can be as simple as, "the graph shows a straight line relation between x and y", or "we can see from the data that y decreases as x increases". This will allow the reader to follow the logic of the readings to the conclusion. Once again, the conclusion drawn has to be **consistent with the data**, address the research question, and will state if the results agree or disagree with the hypothesis. To repeat, the **conclusion does not need to be in line with the scientific consensus**, just in line with your experimental results.

Evaluation (out of 6 points)

The evaluation criteria examinee your ability to critically examine your procedure, data, and findings, and see if there are limitations. They are looking for evidence that you can identify shortcomings to the experiment, caused by design, equipment, or approximations, and how improvements can be made to address these issues. It also looks for evidence of you understanding how your results fit in with the current scientific context.

"A detailed conclusion is described and justified which is entirely relevant to the research question and fully supported by the data presented."

This joins on from the analysis section - your conclusion should reference the research question, and answer it using all the results you have found from your experiments. "*Fully supported*" means that any conclusions you make must use the results as evidence. If you find that you are not able to answer all parts of the research question, explain why this is so; it may be perhaps due to limitations in the availability of equipment, complexity of the problem, or insufficient data. This gives you things to talk about in terms of improvements to the experiment, in the improvements and future investigations sections.

"A conclusion is correctly described and justified through relevant comparison to the accepted scientific context."

The conclusion also needs to be compared with what is the accepted science - this is done either through comparing the values you have obtained with literature values (also sometimes called theoretical values), or discussing if the trends you have observed in your data are in line with scientific concepts. This will be a good time to refer back to the background theory section to explain your results in line with the relevant theory. If there are differences, note these and consider possible reasons for the discrepancy in the evaluation section, considering what other experiments may be done to ascertain the cause of the error in further investigation.

"Strengths and weaknesses of the investigation, such as limitations of the data and sources of error, are discussed and provide evidence of a clear understanding of the methodological issues involved in establishing the conclusion."

The evaluation section will be where most of this is discussed, although it may be worth mentioning this throughout the report. For example, in the methodology for design limitations, or in the results in qualitative observations (e.g., "reaction Z did not produce any observable results due to the chemical being contaminated, so the results have been discarded"). Discuss in the evaluation sources of error, both random and systematic, and to what degree your experiment has compensated for them. Discuss if these have a material effect errors on your calculations, and the resulting precision and accuracy of the experiment. They do not need to have all been corrected for: these errors may have only become evident as you were performing the experiment. Discussions and recognition of these show the critical thinking required.

"The student has discussed realistic and relevant suggestions for the improvement and extension of the investigation."

This criterium is another reason it is good to bring up errors and limitations in experimental design: the possible improvements and further experiments section are ideal for fulfilling this. In the possible improvements section, suggest changes you can make to the experiment based on your list of limitations, to try and correct these limitations. These suggestions can include processes which were not possible due to equipment or time availability. For example, an experiment for enthalpy of combustion using a spirit burner and heating a beaker could be improved by instead using a bomb calorimeter, which may not have been available to you. Remember that the suggestions have to be **realistic** - equipment that would be available perhaps in a different lab would be fine, but suggesting conducting an experiment in space and zero-g might be pushing it a bit!

The further experiments section serves a slightly different purpose from possible improvements. Generally, these are experiments where the scope of the research question has been broadened. For example, you may have investigated the rate of reaction between an acid and metal. Examples of further experiments, or extensions, may be extending the range of concentrations, different acids, different metals, and so forth.

Communication (out of 4 points)

This assessment criteria are based on how you present your report, and how clear the focus, process, and outcomes are. It is important for notation to be consistent, and for the reasoning to follow a natural flow. These will be demonstrated throughout the report, rather than in one particular section.

"The presentation of the investigation is clear. Any errors do not hamper understanding of the focus, process and outcomes."

"The report is well structured and clear: the necessary information on focus, process and outcomes is present and presented in a coherent way."

The reader should be able to have a good idea of what you are trying to do, how you are doing it, and how you have reached your conclusions. This is the main purpose for setting out the IA in the recommended format: use the structure to state the required points that your investigation is focussed on, work through the process, and present the outcome at the end. Note that it says "*any errors do not hamper understanding*" - the IA doesn't have to be perfect for this criteria, just that it has to be clear. Ensure coherence by being clear and **concise** in your writing, and avoiding unnecessary jargon. As mentioned, the report should be accessible to a fellow IB student, so write for that level of understanding and prior understanding.

"The report is relevant and concise thereby facilitating a ready understanding of the focus, process and outcomes of the investigation."

Write about everything that is relevant to your investigation, and **nothing else**. Whether something is relevant or extraneous can be somewhat subjective, but keep asking yourself the question as you write, "is this related to my experiment and its theory"? The structure can keep you on track. This is where spending too much time (more than one paragraph!) writing stories on why the topic is of interest to you in an attempt to score for personal engagement, will have a negative impact on your grade.

"The use of subject specific terminology and conventions is appropriate and correct. Any errors do not hamper understanding."

This criteria requires you to use appropriate and consistent notation and terminology. For example, if you define a value as k in one equation, do not use the term l for the same constant in another. This can easily happen if you are using equations from different texts, especially in related fields such as Physics, Chemistry, and Engineering, where different notation and definitions are often used for the same concepts. Make sure you understand the equations you are referencing, and instead of copy/pasting them into your IA, rewrite the equations using an equation editor in your word processor so that you ensure consistent notation. Once again, note the "*any errors do not hamper...*": it need not be absolutely perfect, but good enough that the reader does not get confused.

5. FINAL NOTES AND TIPS

Choosing Your Research Question

Having now considered both the structure of the report you are aiming for, bear these in mind when deciding on what research question to choose and investigate. Consider how the question would give you opportunity to address the grading criteria of Personal Engagement, Exploration, and Analysis.

Personal engagement as discussed can be displayed by having some personal relevance, so is simple enough to fulfil: remember and state what led you to think of the question. You want to choose a question which will allow you to draw on chemical concepts and theories that are commensurate with the level of the IBDP, and so it may be worth thinking about how you could apply these ideas as you learn them. These could be either in everyday scenarios, or extensions of an experiment described in the course.

You want to be able to demonstrate your data analysis, and as such you'll want to think about what you would choose as your dependent and independent variables, and a system which lends itself to analysis and multiple trials with different conditions as the independent variable. As such, consider what you would vary (independent) and what you would measure (dependent), and how you would use these in your data analysis.

Note that unless you have had to confirm the exact form of your research question with your teacher beforehand, you can adjust your research question based on the type of data your experiment produces so that it is addressed by your results!

Finally, the IA is different from the Extended Essay in that the concepts and theory investigated to not need to be beyond the scope of the course. As such, experiments which do not differ greatly from those you may have done in the course's labs are acceptable: the experiments to not need to be greatly novel.

Typesetting and Graphing in Your Report

An important part of communication is the consistency, flow, and readability of the writing. Ensure that the notation is correct, and that the structure and hierarchy in the report is clear.

Generally, a text editor such as Microsoft Word, or Google Documents will suffice. If you are comfortable with LaTeX, go for it. The main point is to use the platform and lay out the content in such a way that it is easy to read and follow, with clear breakdowns of sections and headers.

As per described in the grading criteria section, you are strongly encouraged to enter the equations and mathematical notation yourself, using the typesetter's equation editor, rather than copy/pasting or screenshotting the equations from a website or text. This is crucial for consistency both in layout and notation.

Plotting graphs can be done in most spreadsheet or data analysis software; the most accessible and available to you may be Microsoft Excel, GSheets, or LibreOffice Calc. These should serve for fitting linear charts, but for more powerful graphing and fitting options are OriginLab, Logger Pro, and CurveExpert.

Keeping the Grading Criteria in Mind

To sum up, the main focusses and purpose of the IA is to encourage you to apply personal investigation, critical thinking and planning, and clear communication to your group 4 subject. The grading criteria are written to reflect this. At each stage of the IA, from planning, through carrying out the experiment and analysis, to writing up the report, try to bear in mind the skills that you are required to display. At each point, think:

- How and why did I decide to do it this way?

- What does this result tell me?

- How does this address my research question?

- What are the implications of the results?

- Does this make sense to me, or with the theory?

As long as you bear these in mind throughout, and follow the structure, you will be able fulfil the criteria and score well on your internal assessment. All the best with your assessments!

PART II

SEVEN EXAMPLES OF EXCELLENT INTERNAL ASSESSMENT

The IAs featured in this section are all recently submitted assessments that scored exceptionally well after being moderated by the IBO. To prevent plagiarism and duplication of results, the appendices have been omitted. The IAs are presented in the exact same way as they were submitted, and without any edits or changes to formatting.

We do not retain the copyright of these commentaries, nor is this publication endorsed by the IBO. The Internal Assessments are being re-printed with the permission of the original authors.

1. THE EFFECT OF INCREASING VOLTAGE ON THE RATE OF ELECTROLYSIS OF COPPER (II) SULFATE

Author: Marco Buttigieg
Moderated Mark: 23/24

The main goal of this investigation was to determine the effect of increasing voltage on the number of moles of hydrogen ions produced in the electrolysis of copper (II) sulfate with inert graphite electrodes. This was done by creating an electrolytic cell and performing the electrolysis at various voltage settings while recording the change in pH of the solution.

Electrolytic cells are a type of electrochemical cell that utilizes electrical energy to drive a non-spontaneous chemical reaction in which electrons are transferred from one chemical species to another. Electrolytic cells consist of the battery or power source, wires, an electrolyte solution, and two electrodes: negative (cathode), where the reduction half-reaction occurs and positive (anode), where the oxidation half-reaction occurs. The two half-reactions combine to form the redox reaction that occurs in the cell (Bylikin et al., 2014). In the electrolysis of copper (II) sulfate with inert graphite electrodes, an electric current is driven through a $CuSO_4$ solution via the graphite electrodes causing the following reactions:

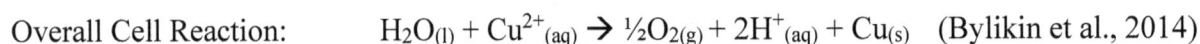

Anode (Oxidation): $\quad\quad\quad\quad H_2O_{(l)} \rightarrow \frac{1}{2}O_{2(g)} + 2H^+_{(aq)} + 2e^-$

Cathode (Reduction): $\quad\quad\quad Cu^{2+}_{(aq)} + 2e^- \rightarrow Cu_{(s)}$

Overall Cell Reaction: $\quad\quad\quad H_2O_{(l)} + Cu^{2+}_{(aq)} \rightarrow \frac{1}{2}O_{2(g)} + 2H^+_{(aq)} + Cu_{(s)}$ $\quad$ (Bylikin et al., 2014)

The rates at which the products of this reaction are produced are dependent on a multitude of factors. The factor that was selected to be the independent variable for this experiment was voltage. Measured in volts, the power source of the electrolytic cell was set to various voltage settings (3V, 6V, 9V, 12V, 15V) to provide the different voltages to the cell.

Voltage is related to current through Ohm's Law, which states that the current (I) of a circuit is equal to voltage (V) divided by resistance (R).

The current of the circuit can be multiplied by the duration of the electrolysis (t) to find the electrical charge (Q) of the circuit. The two relations are as shown below:

$\quad\quad\quad$ Ohm's Law: $I = V / R$ $\quad\quad\quad$ Electrical Charge: $Q = I \times t$ $\quad$ (Milikan and Bishop, 1917)

The charge of a circuit is related to the number of moles of electron flowing through it by Faraday's constant, 96500C / 1mol e^-. According to Faraday's first law of electrolysis, the mass of an element deposited during electrolysis is directly proportional to the charge passing through during the electrolysis (Bylikin et al., 2014).

The dependent variable selected to measure this change is the moles of hydrogen ions produced by the electrolysis. As hydrogen ions are a product of the reaction, their concentration in the electrolyte solution is a good indicator of how far the reaction progressed. At the anode, hydrogen ions were produced via the oxidation of water molecules. First, the pH of the electrolyte solution was measured before and after the electrolysis performed using a Vernier pH probe connected to a computer with Vernier Logger Pro 3.15 software that processes and displays the pH value. The pH, power of hydrogen, is a measure of the hydrogen ion concentration in a solution by the relation:

$$[H^+] = 10^{-pH}$$ (Bylikin et al., 2014)

A hydrogen ion concentration in mol/L was calculated for both before and after. The initial concentration was subtracted from the final concentration to find the concentration change caused by the electrolysis. The concentration change of hydrogen ions was then converted to the number moles of hydrogen ions produced by the electrolysis. The results of the experiment were then analyzed to observe trends in the data and answer the proposed question.

In the summer when I 12 years old, I was trying to get home from the park when I heard thunder, and while walking home I was struck by lightning. Miraculously, I wasn't seriously injured and had a quick recovery in the hospital. Ever since then, anything to do with electricity has intrigued me, as I always sought to fully understand the lightning that hit me on that summer afternoon. Learning about electrochemistry, specifically about electrochemical cells allowed me to connect one of my favourite subjects to the electricity I am so fascinated by. Being able to manipulate a chemical reaction with electricity seemed like it would make for an interesting investigation, so I decided to conduct an electrolysis.

Research Question:

How does changing the voltage in an electrolytic cell affect the production of hydrogen ions in the electrolysis of copper (II) sulfate with inert electrodes?

Hypothesis:

If voltage is increased, then more moles of hydrogen ions will be produced because according to Ohm's law, if voltage increases and resistance stays constant, current will also increase. Increasing current will allow more charge to pass through the circuit, thus producing more moles of hydrogen ions.

Table 1: Controlled Variables

Controlled Variable	Significance of Variable	Method of Control
Time Elapsed	If the electrolysis was left for different durations of time, different amounts of charge would pass through the electrodes, creating varying amounts of product.	Each electrolysis was performed for a duration of $900\pm1s$ (15min).
Ions in Solution	Ions with different ionic charge will require different quantities of electrons for a reaction to occur, producing different amounts of product. Different ions also have different electrode potentials and could take preference over the desired reactants.	Each electrolysis was conducted using a $CuSO_4$ solution, which dissociates into Cu^{2+} and SO_4^{2-} ions.
Solution Concentration	Solutions of varying concentrations will allow different amounts of current to pass through it, ultimately changing the amount of product formed.	Each electrolysis was performed using 1 mol/L $CuSO_4$ solution.
Solution Volume	Differing volumes of solution will contain different numbers of ions in the solution available for reaction, changing the rate of reaction, altering the amount of product produced. Different volumes will also dilute the products to different concentrations, thus making pH measurements inconsistent.	Each electrolysis was conducted with $200\pm1mL$ of $CuSO_4$ solution.
Electrode Composition	Different types of electrodes can be either inert (do not participate in the reaction) or active (participate in the reaction), changing the chemical species produced by the reaction.	Each electrolysis was performed using inert graphite electrodes.
Cell Components Used	Inconsistencies in wiring and in the power source can create discrepancies in the amount of current passing through the apparatus by affecting the resistance of the circuit, ultimately changing the amount of product produced.	Each electrolysis was conducted using the same alligator clips, wiring, ammeter, and power source to maintain a constant electrical resistance
Measuring Equipment Used	Inconsistencies in the measuring equipment used would cause different concentrations of the solution to be created, different currents measured, and differences in measured pH.	The $CuSO_4 \cdot 5H_2O$ solute was measured using the same electric balance for each solution. The same ammeter was used to measure current for all trials. Each solution's pH before and after electrolysis was measured using the same pH probe.

Materials and Apparatus:

- 1248.50g Copper (II) Sulfate Pentahydrate ($CuSO_4 \cdot 5H_2O$) (Ward's Science)
- 4550mL Distilled Water
- 2 Graphite Electrodes
- 4 Alligator Clip Wires
- Elenco Precision Battery Eliminator Model XP-100 (±5%) (Maximum Current - 1A)
- Globe EDM-05 Analog Ammeter (±5mA)
- 500mL Beaker
- Vernier pH Probe (±0.01 pH)
- Vernier Logger Pro 3.15 Software
- Laptop Compatible with Vernier Logger Pro 3.15 Software
- Electronic Balance (±0.01g)
- Weighing Boat
- Glass Stirring Rod
- 200mL Graduated Cylinder (±1mL)
- Timer (±1s)

Procedure:

1. Vernier Logger Pro 3.15 Software was downloaded and installed on the laptop
2. Vernier pH Probe was set up and connected to the computer as stated in the user manual
3. The components of the electrolytic cell were assembled
   i) One end of an alligator clip was attached to the positive (red) output of the battery eliminator, with the other end attached to the top of a graphite electrode that will act as the anode
   ii) One end of an alligator clip was attached to the negative (black) output of the battery eliminator, with the other end attached to the top of a graphite electrode that will act as the cathode
   iii) One end of an alligator clip was attached to the positive (red, 500) terminal of the ammeter
   iv) One end of an alligator clip was attached to the negative (black, -) terminal of the ammeter
   v) The battery eliminator was plugged into a wall outlet
4. 200mL of 1mol/L $CuSO_4$ solution was prepared
   i) 182mL of distilled water was measured in a 200mL graduated cylinder and poured into the 500mL beaker
   ii) 49.94g (0.2mol) of solid $CuSO_4 \cdot 5H_2O$ was measured on an electronic balance using a weighing boat
   iii) The $CuSO_4 \cdot H_2O$ was poured into the beaker of distilled water and stirred with a glass stirring rod until dissolved
5. The pH of the solution before electrolysis was measured and recorded using the Vernier pH probe

6. The graphite electrodes were placed in the beaker of $CuSO_4$ solution
7. The loose ends of both alligator clip wires attached to the ammeter were placed in the beaker of $CuSO_4$ solution
8. The battery eliminator was set to an output of 3V (DC)
9. The battery eliminator was switched on and the timer was set to 15min (900s)
10. The current flowing through the cell was measured and recorded using the ammeter
11. After exactly 15min, the battery eliminator was switched off and the graphite electrodes and the loose ends of the wires connected to the ammeter were removed from the solution
12. The pH of the solution after electrolysis was measured and recorded using the Vernier pH probe
13. Steps 4-12 were repeated for a total of five trials at each of the five voltage settings (3V, 6V, 9V, 12V, 15V)
14. The initial and final pH readings for each trial were converted into hydrogen ion concentrations (mol/L) and the initial was subtracted from the final to find the change in hydrogen ion concentration
15. Change in hydrogen ion concentrations were converted into moles to find the number of moles of hydrogen ions produced by the electrolysis
16. Mean number of moles of hydrogen ions produced were taken for each voltage setting
17. Mean results were plotted on a graph with a trendline
18. Mean current was calculated at each voltage setting in order to determine a theoretical hydrogen ion yield for each voltage, along with a percentage error to compare the experimental and theoretical yield of hydrogen ions

Safety Concerns:

Electric Shock - When working with electrical components, it is necessary to take precautions to mitigate the risk of electrical shock, which can cause a variety of injuries ranging from slight discomfort to cardiac arrest (Hydro Quebec, 2019). During the course of this experiment only wires insulated with rubber were used, and while the power source was switched on and current was flowing through the cell's circuit all contact was avoided with the charged components.

Chemical Irritation – When working with copper (II) sulfate and acidic solutions, it is important to limit chemical contact with the body to avoid injury. $CuSO_4$ is toxic when ingested or inhaled, and is corrosive to the skin and eyes (Fischer Scientific, 2018). Acidic solutions such as the one created by the electrolysis are corrosive to the skin and eyes in more concentrated solutions (Flinn Scientific, 2016). In order to avoid injury, personal protective equipment such as gloves and safety goggles were worn, and the experiment was conducted in a well-ventilated room.

Analysis

Observations:

Table 2: Qualitative Observations of Electrolytic Cell during Electrolysis at Cathode, Anode, and CuSO₄ Solution

Cell Component	Observations
Anode (+ electrode)	Colourless bubbles of oxygen gas formed on electrode → Bubbles produced more vigorously as voltage increased
Cathode (- electrode)	Layer of lustrous reddish-brown copper deposited on electrode → Thickness of deposited copper layer increased with higher voltage
CuSO₄ Solution	Very little visible change in deep blue colour of solution at all voltage settings

Table 3: Raw Current, Initial and Final pH Values for Electrolyte Solutions

Voltage (V) (±5%)	Trial Number	Current (mA) (±5mA)	Initial pH (±0.01)	Final pH (±0.01)
3	1	28	3.68	3.03
	2	28	3.51	3.01
	3	23	3.61	3.08
	4	25	3.72	3.09
	5	31	3.57	2.99
6	1	38	3.74	2.85
	2	40	3.79	2.81
	3	36	3.83	2.88
	4	42	3.56	2.75
	5	39	3.65	2.80
9	1	70	3.76	2.64
	2	68	3.68	2.66
	3	62	3.74	2.74
	4	76	3.62	2.59
	5	64	3.79	2.72
12	1	92	3.68	2.48
	2	81	3.74	2.59
	3	84	3.81	2.57
	4	87	3.68	2.52
	5	86	3.71	2.53
15	1	106	3.75	2.42
	2	112	3.69	2.39
	3	104	3.73	2.45
	4	98	3.62	2.46
	5	95	3.73	2.49

Data Processing:

CuSO₄ Solution Preparation:

$n = MV$

$n = 1.00 \text{ mol/L} \times 0.200\text{L}$

$n = 0.200 \text{ mol } CuSO_4 \cdot 5H_2O \text{ required}$

$0.200\text{mol } CuSO_4 \cdot 5H_2O \left(\frac{249.72\text{g}}{1\text{mol } CuSO_4 \cdot 5H_2O}\right) = \textbf{49.94g } CuSO_4 \cdot 5H_2O \textbf{ required}$

$0.200\text{mol } CuSO_4 \cdot 5H_2O \left(\frac{5\text{mol } H_2O}{1\text{mol } CuSO_4 \cdot 5H_2O}\right)\left(\frac{18.02\text{g}}{1\text{mol } H_2O}\right)\left(\frac{1\text{mL } H_2O}{1\text{g } H_2O}\right) = 18\text{mL } H_2O \text{ in salt}$

$200\text{mL Solution} - 18\text{mL } H_2O \text{ in salt} = \textbf{182mL distilled water required}$

H^+ Concentration (Initial $[H^+]$, 3V, Trial 1):

$[H^+] = 10^{-pH}$

$[H^+] = 10^{-3.68}$

$[H^+] = 2.09 \pm 0.05 \times 10^{-4} \text{ mol/L}$

Error Propagation:

$2.303 \times 0.01 = 0.02303$

$0.2303 \times 2.09 \times 10^{-4} = \textbf{0.05} \times \textbf{10}^{-4} \textbf{ mol/L}$

Change in H^+ Concentration (3V, Trial 1):

$\text{Change } [H^+] = \text{Final } [H^+] - \text{Initial } [H^+]$

$\text{Change } [H^+] = 9.33 \times 10^{-4} \text{ mol/L} - 2.09 \times 10^{-4} \text{ mol/L}$

$\text{Change } [H^+] = 7.24 \pm 0.26 \times 10^{-4} \text{ mol/L}$

Error Propagation:

$0.05 \times 10^{-4} \text{ mol/L} + 0.21 \times 10^{-4} \text{ mol/L} = \textbf{0.26} \times \textbf{10}^{-4} \textbf{ mol/L}$

Moles of H^+ Produced (3V, Trial 1):

$n = MV$

$n = 7.24 \times 10^{-4} \text{ mol/L} \times 0.200\text{L}$

$n = 1.45 \pm 0.05 \times 10^{-4} \text{ mol } H^+$

Error Propagation:

$0.26 \times 10^{-4} \text{ mol/L} \times 0.200\text{L} = \textbf{0.05} \times \textbf{10}^{-4} \textbf{ mol}$

Mean (Moles of H^+ Produced, 3V):

$\bar{x} = \frac{1}{N}\sum_{i=1}^{N} X_i$

$\bar{x} = \frac{1}{5}\sum_{i=1}^{5} X_i$

Error Propagation:

$(0.05 \times 10^{-4} \text{ mol} + 0.06 \times 10^{-4} \text{ mol} + 0.05 \times 10^{-4} \text{ mol} + 0.05 \times 10^{-4} \text{ mol} + 0.05 \times 10^{-4} \text{ mol}) / 5$

$= \textbf{0.05} \times \textbf{10}^{-4} \textbf{ mol}$

$\bar{x} = \frac{1.45 \times 10^{-4} \text{ mol} + 1.34 \times 10^{-4}\text{mol} + 1.17 \times 10^{-4}\text{mol} + 1.24 \times 10^{-4}\text{mol} + 1.50 \times 10^{-4}\text{mol}}{5}$

$\bar{x} = 1.34 \pm 0.05 \times 10^{-4} \text{ mol } H^+$

<u>Theoretical H⁺ Yield (3V):</u>

$Q = It$

$Q = 27mA \left(\frac{1A}{1000mA}\right) \times 900s$

$Q = 24.3 \pm 4.5C$

$24.3 \pm 4.5\ C \left(\frac{1\ mol\ e^-}{96500C}\right) \left(\frac{1\ mol\ H^+}{1\ mol\ e^-}\right) = \mathbf{2.52 \pm 0.47 \times 10^{-4}\ mol\ H^+}$

<u>Percent Error (3V):</u>

$\% \text{ Error} = \left|\frac{Actual\ Yield - Thoeretical\ Yield}{Theoretical\ Yield}\right| \times 100\%$

$\% \text{ Error} = \left|\frac{1.34 \times 10^{-4}\ mol - 2.52 \times 10^{-4}\ mol}{2.52 \times 10^{-4}\ mol}\right| \times 100\%$

% Error = 46.8%

Error Propagation:

$5mA/1000mA = 0.005A$ $(1s/900s) \times 100\% = 0.1\%$

$(0.005/0.0027) \times 100\% = 18.5\%$ $18.5\% + 0.1\% = 18.6\%$

$(18.6\%/100\%) \times 2.52 \times 10^{-4} = \mathbf{0.47 \times 10^{-4}\ mol}$

Table 4: Processed Hydrogen Ion Concentrations and Moles of Hydrogen Produced by Electrolysis

Voltage (V) (±5%)	Trial Number	Initial [H⁺] (mol/L)	Final [H⁺] (mol/L)	Change in [H⁺] (mol/L)	H⁺ Produced (mol)
3	1	$2.09 \pm 0.05 \times 10^{-4}$	$9.33 \pm 0.21 \times 10^{-4}$	$7.24 \pm 0.26 \times 10^{-4}$	$1.45 \pm 0.05 \times 10^{-4}$
	2	$3.09 \pm 0.07 \times 10^{-4}$	$9.77 \pm 0.23 \times 10^{-4}$	$6.68 \pm 0.30 \times 10^{-4}$	$1.34 \pm 0.06 \times 10^{-4}$
	3	$2.45 \pm 0.06 \times 10^{-4}$	$8.32 \pm 0.19 \times 10^{-4}$	$5.87 \pm 0.25 \times 10^{-4}$	$1.17 \pm 0.05 \times 10^{-4}$
	4	$1.91 \pm 0.04 \times 10^{-4}$	$8.13 \pm 0.19 \times 10^{-4}$	$6.22 \pm 0.23 \times 10^{-4}$	$1.24 \pm 0.05 \times 10^{-4}$
	5	$2.69 \pm 0.06 \times 10^{-4}$	$1.02 \pm 0.02 \times 10^{-3}$	$7.51 \pm 0.26 \times 10^{-4}$	$1.50 \pm 0.05 \times 10^{-4}$
6	1	$1.82 \pm 0.04 \times 10^{-4}$	$1.41 \pm 0.03 \times 10^{-3}$	$1.23 \pm 0.03 \times 10^{-3}$	$2.46 \pm 0.06 \times 10^{-4}$
	2	$1.62 \pm 0.04 \times 10^{-4}$	$1.55 \pm 0.04 \times 10^{-3}$	$1.39 \pm 0.04 \times 10^{-3}$	$2.78 \pm 0.08 \times 10^{-4}$
	3	$1.48 \pm 0.03 \times 10^{-4}$	$1.32 \pm 0.03 \times 10^{-3}$	$1.17 \pm 0.03 \times 10^{-3}$	$2.34 \pm 0.06 \times 10^{-4}$
	4	$2.75 \pm 0.06 \times 10^{-4}$	$1.78 \pm 0.04 \times 10^{-3}$	$1.50 \pm 0.05 \times 10^{-3}$	$3.01 \pm 0.10 \times 10^{-4}$
	5	$2.24 \pm 0.05 \times 10^{-4}$	$1.58 \pm 0.04 \times 10^{-3}$	$1.36 \pm 0.05 \times 10^{-3}$	$2.72 \pm 0.10 \times 10^{-4}$
9	1	$1.74 \pm 0.04 \times 10^{-4}$	$2.29 \pm 0.05 \times 10^{-3}$	$2.12 \pm 0.05 \times 10^{-3}$	$4.23 \pm 0.10 \times 10^{-4}$
	2	$2.09 \pm 0.05 \times 10^{-4}$	$2.19 \pm 0.05 \times 10^{-3}$	$1.98 \pm 0.06 \times 10^{-3}$	$3.96 \pm 0.12 \times 10^{-4}$
	3	$1.82 \pm 0.04 \times 10^{-4}$	$1.82 \pm 0.04 \times 10^{-3}$	$1.64 \pm 0.04 \times 10^{-3}$	$3.28 \pm 0.08 \times 10^{-4}$
	4	$2.40 \pm 0.06 \times 10^{-4}$	$2.57 \pm 0.06 \times 10^{-3}$	$2.33 \pm 0.07 \times 10^{-3}$	$4.66 \pm 0.14 \times 10^{-4}$
	5	$1.62 \pm 0.04 \times 10^{-4}$	$1.91 \pm 0.04 \times 10^{-3}$	$1.74 \pm 0.04 \times 10^{-3}$	$3.49 \pm 0.08 \times 10^{-4}$
12	1	$2.09 \pm 0.05 \times 10^{-4}$	$3.31 \pm 0.08 \times 10^{-3}$	$3.10 \pm 0.09 \times 10^{-3}$	$6.20 \pm 0.18 \times 10^{-4}$
	2	$1.82 \pm 0.04 \times 10^{-4}$	$2.57 \pm 0.06 \times 10^{-3}$	$2.39 \pm 0.06 \times 10^{-3}$	$4.78 \pm 0.12 \times 10^{-4}$
	3	$1.55 \pm 0.04 \times 10^{-4}$	$2.69 \pm 0.06 \times 10^{-3}$	$2.54 \pm 0.06 \times 10^{-3}$	$5.08 \pm 0.12 \times 10^{-4}$
	4	$2.09 \pm 0.05 \times 10^{-4}$	$3.02 \pm 0.07 \times 10^{-3}$	$2.81 \pm 0.08 \times 10^{-3}$	$5.62 \pm 0.16 \times 10^{-4}$
	5	$1.95 \pm 0.04 \times 10^{-4}$	$2.95 \pm 0.07 \times 10^{-3}$	$2.76 \pm 0.07 \times 10^{-3}$	$5.52 \pm 0.14 \times 10^{-4}$
15	1	$1.78 \pm 0.04 \times 10^{-4}$	$3.80 \pm 0.09 \times 10^{-3}$	$3.62 \pm 0.09 \times 10^{-3}$	$7.24 \pm 0.18 \times 10^{-4}$
	2	$2.04 \pm 0.05 \times 10^{-4}$	$4.07 \pm 0.09 \times 10^{-3}$	$3.87 \pm 0.10 \times 10^{-3}$	$7.74 \pm 0.20 \times 10^{-4}$
	3	$1.86 \pm 0.04 \times 10^{-4}$	$3.55 \pm 0.08 \times 10^{-3}$	$3.36 \pm 0.08 \times 10^{-3}$	$6.72 \pm 0.16 \times 10^{-4}$
	4	$2.40 \pm 0.06 \times 10^{-4}$	$3.47 \pm 0.08 \times 10^{-3}$	$3.23 \pm 0.09 \times 10^{-3}$	$6.46 \pm 0.18 \times 10^{-4}$

	5	$1.86 \pm 0.04 \times 10^{-4}$	$3.24 \pm 0.07 \times 10^{-3}$	$3.05 \pm 0.07 \times 10^{-3}$	$6.10 \pm 0.14 \times 10^{-4}$

Table 5: Mean Moles of Hydrogen Ion Produced and Current with Theoretical Hydrogen Ion Yield

Voltage (V) (±5%)	Mean H^+ Produced (mol)	Mean Current (mA) (±5mA)	Theoretical Yield of H^+ (mol)	Percent Error (%)
3	$1.34 \pm 0.05 \times 10^{-4}$	27	$2.52 \pm 0.47 \times 10^{-4}$	46.8
6	$2.66 \pm 0.08 \times 10^{-4}$	39	$3.64 \pm 0.47 \times 10^{-4}$	26.9
9	$3.92 \pm 0.10 \times 10^{-4}$	68	$6.34 \pm 0.48 \times 10^{-4}$	38.2
12	$5.44 \pm 0.14 \times 10^{-4}$	86	$8.02 \pm 0.47 \times 10^{-4}$	32.2
15	$6.85 \pm 0.17 \times 10^{-4}$	103	$9.61 \pm 0.48 \times 10^{-4}$	28.7

Figure 1: Moles of H^+ Produced During Electrolysis at Various Voltage Settings

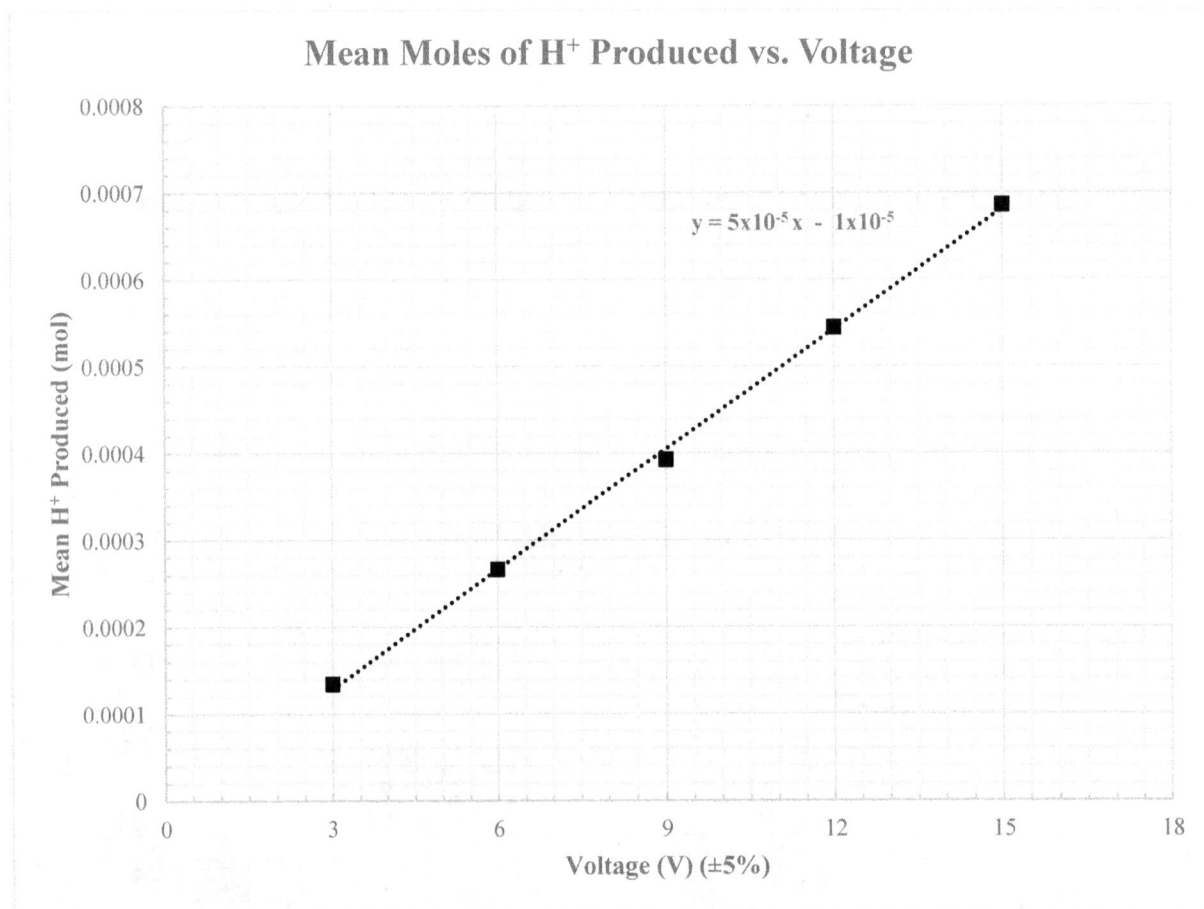

Mean Moles of H^+ Produced vs. Voltage

$y = 5 \times 10^{-5} x - 1 \times 10^{-5}$

Evaluation

After the experiment was concluded and results were analyzed, a strong trend was discovered in the data, relating voltage to the amount of product formed by the electrolysis of copper (II) sulfate using inert electrodes, seen through the number of moles of hydrogen ions produced by the electrolysis. At the lowest voltage, 3V, there was the smallest number of moles of hydrogen ions produced, with a mean value of just 1.34×10^{-4} mol. When voltage was the highest at 15V, the greatest production of hydrogen ions was also observed, with a mean of 6.85×10^{-4} mol produced. As voltages increased within this range, a very strong positive linear correlation was observed between voltage and moles of hydrogen ions produced, with all data points falling on or very close to the trendline. Thus, the data supports the initial hypothesis, showing that as voltage increases, the number of moles of hydrogen ions produced by the electrolysis also increases.

This was the expected result of the experiment, agreeing with the proposed scientific theory. By Ohm's law, an increasing voltage at a constant resistance will cause an increase in electrical current (Milikan and Bishop, 1917). Similarly, increasing current at a constant time will generate an increased electrical charge. According to Faraday's first law of electrolysis, the mass of an element deposited during electrolysis is directly proportional to the charge passing through during the electrolysis (Bylikin et al., 2014). The theory points towards a proportional increase in products created by increasing voltage.

The data collected also generally follows this trend, as each time the voltage doubles, the moles of hydrogen ions similarly doubled. For example, from 3V to 6V, the voltage is doubled. The moles of hydrogen ions produced also doubles from $1.34 \pm 0.05 \times 10^{-4}$ mol to $2.66 \pm 0.08 \times 10^{-4}$ mol. Though not exactly twice the value, a true doubled value falls within the uncertainty, so the discrepancy can be attributed to random error. Thus, the increase seen in the experiment is also proportional.

To further test the reliability of the data, a theoretical value for the number of moles of hydrogen ions produced was calculated using the average electrical current at each voltage. The theoretical and experimental yields of hydrogen ions were compared via percentage error. With errors ranging from 26.9% to 46.8%, random error alone cannot account for this difference, so systematic error must have also had an impact on the experiment.

The first systematic error should be noted is that the theoretical values used are not truly theoretical, and instead, are based upon experimental current values flowing through the circuit. This was done because there was no means of determining the resistance of the circuit to calculate an expected electrical current via Ohm's law. Due to this, the true theoretical yield of this experiment could not be determined, so the calculated percentage error is not entirely accurate. This could have been avoided if electrical resistance was measured using an ohmmeter, then using Ohm's law to calculate a theoretical current.

The measured electrical current during experimentation also provides a source of systematic error. Theoretically, if the components and wiring of the cell were identical during each trial, every trial at the same voltage should have also had the same amount of electrical current. Since the current varied slightly between trials, it can be assumed that there were some changes in circuit resistance. The variances in current would have created different amounts of charge in the circuit, thus affecting the number of moles of hydrogen ions produced. This error could have been reduced by using a hardwired cell rather than alligator clip wiring, as this would have made the contact between cell components more consistent, helping to maintain a constant resistance.

A final source of systematic error was exhibited by the differences in the initial pH of the electrolyte solution. All electrolyte solutions were prepared to be 200mL of 1mol/L $CuSO_4$. Identical solutions should theoretically also have very similar initial pH values. This indicates that there were some inconsistencies in the electrolyte solution related to the concentration of $CuSO_4$. $CuSO_4$ is an acidic salt, so higher concentrations would lead to a lower solution pH. Having slightly different concentrations of $CuSO_4$ in solution would directly affect the number of ions in the solution. Having different quantities of ions in the electrolyte solution impacts the amount of current that can pass through it, ultimately impacting the number of moles of hydrogen ion produced by the electrolysis. This could have been avoided by preparing a 5L, 1 mol/L $CuSO_4$ solution, so all trials would use some of the same solution, ensuring consistency in $CuSO_4$ concentration.

A change that could have been made to improve the experiment would have been to conduct several more trials, perhaps 10-15 per voltage setting in order to increase the precision of the experiment. Additionally, adding a larger range of voltages would have allowed a wider variety of electrical currents to flow through the circuit would find if the trend found in this investigation applies to a larger set of voltages. It would also determine if there is a cap on the amount of current that can flow and thus an upper limit of products that can be produced via the electrolytic cell. Further experimentation should also incorporate other redox reactions with different electrolytic solutions or active electrodes to ensure that the trend can be observed over all redox reactions.

References

Bylikin, S., Horner, G., Murphy, B., Tarcy, D. (2014). *Chemistry Course Companion.* Oxford: Oxford University Press.

Fischer Scientific. (2018). *Copper (II) Sulfate Safety Data Sheet.* Retrieved February 11, 2019, from https://www.fishersci.com/msds?productName=AC422870050

Flinn Scientific. (2016). *Acid Safety: Safety Tips for Using Acids in School Laboratories.* Retrieved February 11, 2019, from https://www.blueridge.edu/sites/default/files/pdf/ehsi/Acid Safety.pdf

Hydro Quebec. (2019). *The Possible Consequences of an Electric Shock on the Body*. Retrieved February 11, 2019, from https://www.hydroquebec.com/safety/electric-shock/consequences-electric-shock.html

2. TO DETERMINE THE EFFECT OF INCREASING TEMPERATURE ON DISSOLVED OXYGEN IN TAP WATER USING WINKLER TITRATION METHOD.

Author: Shreyansh Jain
Moderated Mark: 23/24

Research Question: To determine the effect of increasing temperature on dissolved oxygen in tap water using Winkler Titration method.

Introduction

Global warming in the present time has taken temperature to the highest level. This rise in temperature had led to high mortality rate of organisms. I possess a major interest in Ecology so I frequently read articles related with these issues. A current article I read underlined the concern of increasing mortality rate of salmon due to extreme increase in temperature of water. The article highlighted the deaths of various Alaskan salmon. It also stated that around 850 unspawned salmon died due to heat stress[3]. Well, I was familiar with the fact that the metabolic reaction within the living body is affected by the temperature. Although, the increase in temperature was the reason behind the mass deaths but I was willing to know about the other factors also. Later in the article it was clearly mentioned that "Physiologically, the fish can't get oxygen moving through their bellies", and when I researched upon it, I discovered that one of the major factors was the depletion of oxygen level in water due to high temperature. This increased my curiosity level of determining the relation between dissolved oxygen level and temperature.

Background Information

Dissolved oxygen is the amount of free oxygen present in water which is accessible to living aquatic species.[4] Water is a V-shaped molecule in which oxygen is the central atom joined (bonded) with two hydrogen atoms. Due to the difference in electronegativities of hydrogen and oxygen atoms it is a polar covalent molecule. Partial negative charge ($\delta-$) is present on the oxygen atom because of its higher electronegativity while the partial positive charge ($\delta+$) is present on the hydrogen atom.

[5]

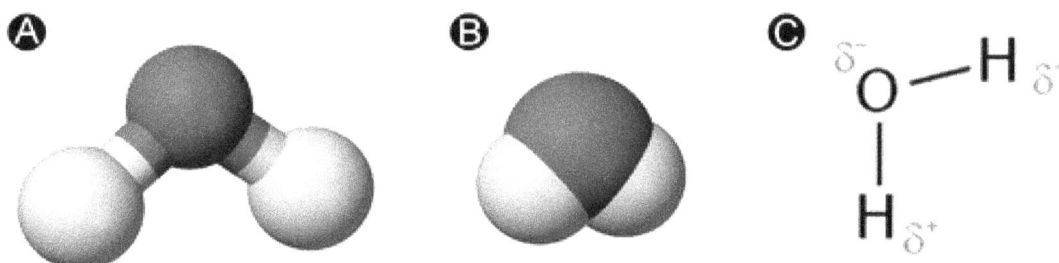

[3] https://edition.cnn.com/2019/08/16/us/alaska-salmon-hot-water-trnd/index.html

[4] https://www.lenntech.com/why_the_oxygen_dissolved_is_important.htm

[5] https://chem.libretexts.org/Bookshelves/Introductory_Chemistry/Book%3A_Introductory_Chemistry_(CK-12)/15%3A_Water/15.01%3A_Structure_of_Water

Figure 1. Molecule of water visualized in 3 distinct ways: (A) ball-and-stick model (B) space-filling model and (C) structural formula with partial charges.

Winkler Method.

Winkler method calculates the Biochemical Oxygen Demand (BOD) in water by redox titrations. The following sequence of redox reactions is based upon this method.[6]

1. Initially $MnSO_4$ is added to fix the dissolved oxygen in water and after the reaction in basic medium, Mn(II) oxidizes into Mn(IV).

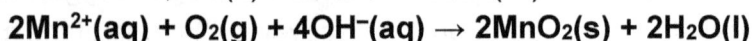

$$2Mn^{2+}(aq) + O_2(g) + 4OH^-(aq) \rightarrow 2MnO_2(s) + 2H_2O(l)$$

2. Mn(IV) oxidizes iodine ion (I^-) into I_2 in acidic medium.

$$2MnO_2(s) + 4I^-(aq) + 8H^+(aq) \rightarrow 2Mn^{2+}(aq) + 2I_2(aq) + 4H_2O(l)$$

3. The iodine which is produced is titrated with sodium thiosulfate.

$$4S_2O_3^{2-}(aq) + 2I_2(aq) \rightarrow 4I^-(aq) + 2S_4O_6^{2-}(aq)$$

In these sequences of reactions, we can see that for every **1 mole of O_2** in the water, **4 mol of $S_2O_3^{2-}$ are used**.

Variables

Independent variable: $250cm^3$ tap-water temperature

- Electrical water maintains the temperature at 35ºC, 45ºC, 55ºC, 65ºC and 75ºC and is checked via thermometer before adding the reactants.
- The experiment was carried out in an air-conditioned room to maintain the constant room temperature and was measured by thermometer in middle of the experiment.

Dependent variable: Amount of dissolved oxygen in tap water

- It will be calculated by Winkler's method.
- We can calculate the amount of oxygen in tap water by making use of the reacted volume of $Na_2S_2O_3.5H_2O$ (hydrated sodium thiosulfate) in titration.

Controlled Variables

---

[6] https://pdfs.semanticscholar.org/8a69/46d53088e4c66086c48ef8ff9c8451dbd2f4.pdf

Controlled Variables	How can we control it?	Why should we control it?
The time for which the samples of water were being heated	The water samples were put into the water bath for 300 seconds at the selected temperatures and the time was determined by a stopwatch	Dissolved oxygen in water samples may get affected when provided with different heating time.
Volume of Alkaline KI and $MnSO_4$ added	$2cm^3$ of 0.6 $moldm^{-3}$ alkaline KI and $2cm^3$ of 0.6 $moldm^{-3}$ $MnSO_4$ and was been added to the samples.	If there is any difference in the volume of alkaline KI and $MnSO_4$ then we cannot predict the number of moles of iodine consumed in the process of titration. Due to this we will not be able to determine oxygen content in the samples by stoichiometric coefficients.
Controlled Variables	How can we control it?	Why should we control it?
Molarity of Alkaline KI and $MnSO_4$ added	0.6 $moldm^{-3}$ alkaline KI, 0.6 $moldm^{-3}$ $MnSO_4$ and 0.6 $moldm^{-3}$ KOH were being used.	Any difference in the concentration of Alkaline KI and $MnSO_4$ would lead to identical results as difference in their volume.
Introduction of concentrated sulfuric acid	Using 2 cm^3 Pasteur glass pipette, 2 cm^3 concentrated sulfuric acid has been added.	Oxygen concentration could be affected by the difference in total volume, in order to avoid any difference, the similar volume has been used across each sample.
Addition of starch to the water samples.	1.2g of starch was added into the solution.	To maintain the same color intensity of solution to prevent differences in color subjectivity.
Amount of water being used	$250.0cm^3$ of water is being used and was measured by measuring cylinder.	If water of different volumes were taken into consideration then it could impact the count of dissolved oxygen moles in water & further influence its concentration. Apart from this, it could also impact the amount of heat being absorbed by distinct water samples.
Samples of water that are being used	Tap water is being taken into consideration.	Tap water samples are being used to know the different concentrations of dissolved oxygen content.
Environmental Pressure	Experiment was conducted under constant pressure of 100kPa with fixed room temperature being set by an air conditioner and also by making sure that the doors and windows of the labs are closed.	As the pressure increases, oxygen solubility also increases So, the pressure was kept constant to eliminate factors apart from temperature influencing oxygen solubility in water.

Method

The method is taken from the method which has been applied by the University of Idaho and Montana State University. Some changes have been made to overcome certain problems.

- We used Erlenmeyer flasks of $250cm^3$ because of lack of availability of BOD bottles (Erlenmeyer flasks are been considered as the best alternatives of BOD bottles). Most of the time the flasks were kept closed during the experiment.
- Reagent concentrations were changed to 0.6 $moldm^{-3}$ alkaline KI, 0.6 $moldm^{-3}$ $MnSO_4$ and 0.06 $moldm^{-3}$ sodium thiosulfate. These concentrations gave the most suitable results, as many other concentrations failed to give the desired changes in the final value after titration and color.
- As an original reagent, in place of alkali iodide-azide, alkali potassium iodide was used.

Preparing Standard Solutions and setting apparatus for titration

1. On a chemical balance, measure 25.33g of hydrated manganese sulfate ($MnSO_4.H_2O$)
2. Transfer this weighted manganese sulfate into $250cm^3$ volumetric flask through a funnel and wash it with distilled water and fill the flask to the mark with distilled water.
3. Make sure that manganese sulfate is fully dissolved. This is 0.6 $moldm^{-3}$ standard solution of $MnSO_4$ (manganese sulfate).
4. Repeat the steps from 1 to 3 with 24.9g KI (potassium iodide) and 8.4g of KOH (potassium hydroxide) added to a different volumetric flask to prepare $0.6 moldm^{-3}$ of KI and KOH respectively.
5. Repeat the steps from 1 to 3 for 3.72g of hydrated sodium thiosulfate ($Na_2S_2O_3.5H_2O$) in order to form 0.06 $moldm^{-3}$ sodium thiosulfate solution.
6. For titration, set $50cm^3$ burette on clamp stand.
7. Wash the burette with $Na_2S_2O_3$ (sodium thiosulfate) and fill it with 0.06 $moldm^{-3}$ of sodium thiosulfate solution.

Heating water samples

8. Using a measuring cylinder, take $250cm^3$ tap water and transfer it in the Erlenmeyer's flask. Don't forget to close the lid.
9. Half fill the electrical water bath with the water sample (tap water) and set the temperature to 25°C. Allow the water to heat to the set temperature and use a thermometer to monitor the temperature.
10. Place the flask of water sample in water bath for 300 seconds.
11. Use heat resistant gloves to take out the flask after set time i.e., 300 seconds.

Figure 2: $250cm^3$ of water sample

12. Transfer $2cm^3$ of manganese sulfate by graduated pipette ($10cm^3$) to the water sample. Make sure that air does not enter while transferring.
13. Repeat the step 12 with $2cm^3$ of alkali potassium iodide.
14. After closing the flask, shake the mixture thoroughly until brown precipitates appears.
15. Add $2cm^3$ concentrated sulfuric acid through a pipette. Ensure, no oxygen gets introduced in the sample. Shake the flask so that the precipitate dissolves.

Titration

16. Keep a glazed tile under the burette and place the Erlenmeyer flask containing the solution above the tile.
17. Note down the initial burette reading and run-down sodium thiosulfate drop by drop.
18. Measure 1.2g of starch on a balance and add it.
19. Shake the flask until solution changes to blue-black color as shown in figure 3.
20. Keep on adding sodium thiosulfate drop by drop through burette until it gets decolorized.
21. Note the final reading of the burette.
22. Repeat the steps from 8 to 21 with 45°C, 55°C, 65°C, 75°C and repeat the experiment 2 more times for each temperature in order to get 3 results.

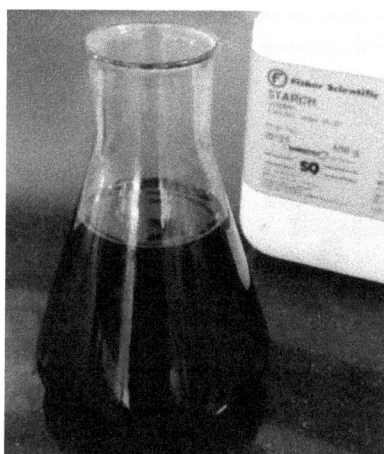

Figure 3: Sample after adding 1.2g of starch

Figure 4: End point of titration

Safety issues

➢ As stated by Nyrstar (safety data sheet, 2017), concentrated sulfuric acid is extremely dangerous and corrosive and can cause severe skin burns and eye damage. Therefore, gloves and glasses must be worn while handling sulfuric acid. In case of any contact with skin, immediately take off the contaminated clothes, wipe the acid with a clean dry cloth and wash it with water.[7]
➢ While using the water bath for heating the flasks, make sure you don't touch the water bath or the water in it with bare hands as it may be hot. Use heat resistant gloves when taking out the flasks from the water bath.

Ethical issue

➢ Plenty of water is consumed in the experiment which could have been used for other purpose. To tackle this issue, perform the experiment with a lesser amount (volume) of water.

---

[7] https://www.nyrstar.com/~/media/Files/N/Nyrstar/sustainability/safety-data-sheets-australia-english/Sulphuric%20Acid.pdf

Environmental issue

> Lot of energy was consumed by electrical water bath in the experiment so to handle the environmental issue, ensure that the water bath has been turned off when not in the use.

Data

Table 1. Measurements of the controlled variables.

Concentration of standard $MnSO_4$ solution / $moldm^{-3}$ (±0)	0.60
Concentration of standard alkaline KI solution / $moldm^{-3}$ (±0)	0.60
Concentration of standard sodium thiosulfate solution / $moldm^{-3}$ (±0)	0.06
Volume of water samples i.e., tap water / cm^3 (±2.5)	250.00
Quantity of sulfuric acid added / cm^3 (±0)*	2.00
Amount of starch / g (±0.001)	1.20
Time for which the samples were heated / s (±0.1)	300.00

Note: Instead of mass or the volume of these solutions, the concentration of standard solutions made are given in table 1 for manganese sulfate, alkaline potassium iodide and sodium thiosulfate. It is assumed that the concentrations of all the standard solutions have the uncertainty of **±0 $moldm^{-3}$**.

*The pipette used for the sulfuric acid has an unknown uncertainty so it is assumed as **±0**.

Table 2. Data of titration.

Temperature of the sample /°C (±0.1)	Volume of titrant (sodium thiosulfate) / cm^3 (±0.05)					
	Trial 1		Trial 2		Trial 3	
	Initial reading	Final reading	Initial reading	Final reading	Initial reading	Final reading
25.0	11.10	14.70	34.50	38.50	18.20	22.00
35.0	7.30	10.00	18.20	21.90	24.00	27.60
45.0	9.50	13.10	21.20	24.70	28.50	32.30
55.0	24.10	27.70	25.70	29.10	37.50	41.00
65.0	27.20	30.50	6.30	10.00	17.80	21.10
75.0	42.00	44.80	44.70	47.80	1.00	3.90

Qualitative observations

• While preparing the standard solutions, the flask felt hot, indicating loss of heat along with exothermic reactions taking place.

- Introduction of alkaline potassium iodide and manganese sulfate into water sample turned it brown and the precipitate were formed when kept shaking it.
- Addition of concentrated sulfuric acid turned the solution into a clear brown color as the precipitate were dissolved.
- Titrating with sodium thiosulfate turned the solution to yellow-brown color & addition of starch changed the solution to blue-black color.
- It was difficult to identify whether the color of the solution has changed or not during the experiment as starch was insoluble in solution.
- It was a slow reaction and spotting the end point was difficult.

Table 3. Data table for calculating the moles of sodium thiosulfate.

Temperature of the sample / °C (± 0.1)	Volume of 0.06 moldm^{-3} Na$_2$S$_2$O$_3$ used as titre / cm^3 (± 0.10)			Average volume of 0.06moldm^{-3} Na$_2$S$_2$O$_3$ used / cm^3 (± 0.10)	Moles of Na$_2$S$_2$O$_3$ ($\times 10^{-4}$) / mol
	Trial 1	Trial 2	Trial 3		
25.0	3.60	4.00	3.80	3.80	2.28
35.0	2.70	3.70	3.60	3.65*	2.19
45.0	3.60	3.50	3.80	3.63	2.18
55.0	3.60	3.40	3.50	3.50	2.10
65.0	3.40	3.70	3.20	3.30*	1.98
75.0	2.80	3.10	2.90	2.93	1.76

*Note: On averaging, the non-concordant values are avoided. The values which differ 0.20cm^3 from each other are known as concordant values.

Sample calculation

> The weight required to prepare standard solutions of $MnSO_4$, KI, $Na_2S_2O_3$ and KOH were estimated by using their molar mass.
>
> $$Molarity = \frac{Mass \times 1000}{Molar\ mass \times volume\ (cm^3)}$$
>
> For example: Calculations for $MnSO_4$
>
> To prepare $0.6 moldm^{-3}$ $MnSO_4$ when hydrated has the molar mass of 168.90.
>
> $$0.6 = \frac{Mass \times 1000}{168.90 \times 250}, \quad Mass = 25.33g$$

Note: All examples are done for trial at room temperature i.e., **25°C**

1. Volume of $0.06 moldm^{-3}$ sodium thiosulfate ($Na_2S_2O_3$) used as titre (trial 1 at 25°C).
   Volume of titre = final reading – initial reading
   $$= 14.70 - 11.10$$
   $$= 3.60cm^3$$

2. Average volume of $0.06 moldm^{-3}$ sodium thiosulfate ($Na_2S_2O_3$) used as titre:
   $$Average\ volume = \frac{Sum\ of\ concordant\ volumes\ in\ all\ trials}{Number\ of\ trials\ with\ concordant\ results}$$
   $$= \frac{3.60 + 4.00 + 3.80}{3} = 3.80cm^3$$

3. Count of moles of sodium thiosulfate ($Na_2S_2O_3$)
   Number of moles = concentration × volume in dm^3
   Concentration of $Na_2S_2O_3$ = $0.06 moldm^{-3}$
   Moles = $0.06 moldm^{-3} \times 0.00380 dm^3$
   $$= 0.000228 mols = 2.28 \times 10^{-4} mols$$

4. Number of moles of oxygen
   As mentioned earlier (background information), when $MnSO_4$ reacts with 1 mole of O_2 in water, 4 moles of $S_2O_3^{2-}$ are used.
   $$\frac{0.000228}{4} = 5.70 \times 10^{-5}$$

Table 4. Data table for calculating the dissolved oxygen concentration in ppm at different temperatures.

Temperature / °C	Moles of $Na_2S_2O_3$ ($\times 10^{-4}$) / mol	Moles of oxygen ($\times 10^{-5}$) / mol	Mass of oxygen in water / g	Concentration of dissolved oxygen / ppm
25	2.28	5.70	0.001824	7.30
35	2.19	5.50	0.001752	7.01
45	2.18	5.40	0.001742	6.97
55	2.10	5.30	0.001680	6.72
65	1.98	5.00	0.001584	6.34
75	1.76	4.40	0.001406	5.62

5. Oxygen mass in 250cm³ water sample

   Mass = moles × molecular mass

   Molecular mass of O_2 = 32.00 gmol⁻¹ (data taken from IB data booklet)

   So, mass of O_2 = 5.70 ×10⁻⁵ × 32.00

   = 1.824 ×10⁻³ g

6. Dissolved oxygen concentration in ppm

   Mass of O_2 in g = 1.824 ×10⁻³ g

   Mass of solute in g per 1cm³ of solution is known as parts per million

   $$\text{Concentration} = \frac{1.824 \times 10^{-3} \times 10^6}{250 cm^3} = 7.30 \text{ ppm}$$

Table 5. Table of Uncertainty

Temperature / °C (±0.1)	% uncertainty / %	Volume of water samples /cm³ (±2.5)	% uncertainty / %	Volume of $MnSO_4$ / cm³ (±0.05)	% uncertainty / %	Volume of Alkaline KI / cm³ (±0.05)	% uncertainty / %	Average volume of $Na_2S_2O_3$ / cm³ (±0.10)	% uncertainty / %	Total % uncertainty / %
25	0.40	250.0	1.0	2.00	3	2.00	3	3.80	3	10
35	0.29	250.0	1.0	2.00	3	2.00	3	3.65	3	10
45	0.22	250.0	1.0	2.00	3	2.00	3	3.63	3	10
55	0.18	250.0	1.0	2.00	3	2.00	3	3.50	3	10
65	0.15	250.0	1.0	2.00	3	2.00	3	3.30	3	10
75	0.13	250.0	1.0	2.00	3	2.00	3	2.93	3	10

If there is uncertainty > 2 then the percentage uncertainties are rounded to 1 s.f and if there is uncertainty is < 2 then it is rounded to 2 s.f.

% Uncertainty = $\frac{Absolute\ Uncertainty}{Measurement\ taken} \times 100$

For 25°C, Percentage Uncertainty = $\frac{0.1}{25} \times 100 = 0.40$

The overall % uncertainty is determined by adding all of the individual percentage uncertainties. For example, for 25°C, the total % uncertainty = 0.40+1+3+3 = 10.40

Since the total % uncertainty is greater than 2, it is rounded to 10 (1s.f)

To calculate the total absolute uncertainty as shown in **table 6**, the oxygen concentration in the sample is multiplied with the total % uncertainty of each temperature and then divide it by 100.

For 25°C, absolute uncertainty = $\frac{10}{100} \times 7.30 = 0.730$

Table 6. Concentration of dissolved oxygen with temperature.

Temperature / °C (±0.1)	Dissolved oxygen concentration / ppm
25	7.30±0.730
35	7.01±0.701
45	6.97±0.697
55	6.72±0.672
65	6.34±0.634
75	5.62±0.562

Graph 1. Relationship between temperature and concentration of dissolved oxygen

Relationship between temperature and concetration of dissolved oxygen

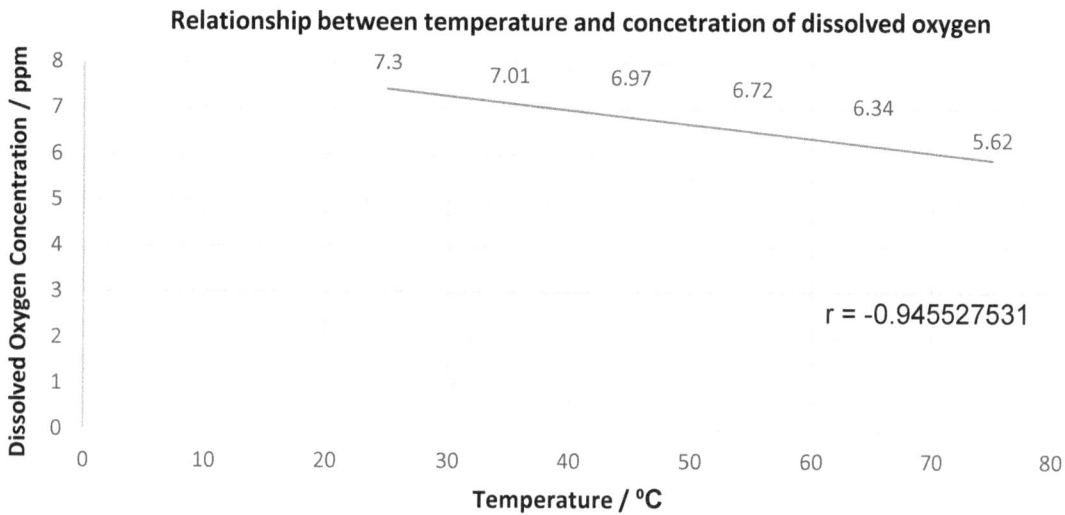

Dissolved Oxygen Concentration / ppm

7.3   7.01   6.97   6.72   6.34   5.62

r = -0.945527531

Temperature / °C

Analysis of the graph

The graph above shows that the dissolved oxygen level in water decreases ass there is an increase in temperature. The calculated value of Pearson's correlation coefficient r is -0.945527531 which determines a very strong negative correlation between the two variables. The graph shows no outliers which makes my result more accurate.

Conclusion

The concentration of dissolved oxygen at 25°C (room temperature) is 7.30 ppm and it falls to 5.62 ppm as we raise the temperature to 75°C by heating it up. Well, this supports my hypothesis and my explanation as well. The results of the experiment seem to be authentic and reliable as the value of Pearson's Correlation Coefficient **r** calculated is -0.94552, which is very close to -1, suggesting that there is a strong negative correlation between the two variables. Thus, I'm quite confident with my results.

Moving on, the % uncertainty calculated is 10% which is acceptable. Due to some methodological error and limited count of repeats for the experiment, the uncertainties are present (see limitations and improvements).

In the opinion of Environmental Protection Agency [Archive.epa.gov, 2012], theoretical value of concentration of dissolved oxygen in water is 8.24 ppm at 25°C, while it is 7.68 ppm at 25°C in my experiment. This suggests that the results of my experiment are a bit different from the theoretical value, hence indicating unreliability of the results to an extent.

Calculation for % error

$$\% \text{ error} = \frac{theoretical\ value - experimental\ value}{theoretical\ value} \times 100$$

$$= \frac{8.24 - 7.30}{8.24} \times 100$$

$$= 11.4\%$$

The percentage error calculated for the experiment is 11.4%. Since my % error is greater than % uncertainty, the main types of error discovered in the experiment are systematic error. These are originated by the loss of heat during the experiment and the use of Erlenmeyer's flask instead of BOD bottles While taking all limitations into the account, the % error is considered to be reasonable. The uncertainties of the experiment would have influenced the results of the experiment. Although, apparatus with high precision ($10cm^3$ pipette and $50cm^3$ burette) were used for the experiment so that the effect of the uncertainties could be minimized. Also, I'm satisfied and assured with my results.

To sum up, the results of the experiment shows that the oxygen level in water decreases as the temperature increases which is one of the reasons for the extinction of Alaskan salmon and many other aquatic species. This increase in temperature also contributes to global warming at a large scale. This investigation therefore signifies that the rise in atmospheric temperature must be controlled so that aquatic species have sufficient access to oxygen.

Strengths

The strengths are controlling the controlled variables to the maximum limit. Also, using the precise apparatus (pipette & burette) for the experiment in order to eliminate the random errors and high uncertainties. Well, one more strength could be the selection of the range of temperature for the experiment in order to answer the research question within the context.

Limitations and improvements

Limitations	Significance	Improvements
Repetition of experiment for each temperature	High significance: The experiment was performed only 3 times; the random errors	Increase the number of experiments with more

	could have been decreased by increasing the number of repeats of experiment.	concordant results for every temperature.
Replacement of BOD bottles	High significance: Erlenmeyer's flasks were used as a substitute of BOD bottles which might stretch the possibility of introducing the oxygen into the sample.	Take use of BOD bottles in titration.
Oxygen control	Oxygen might have entered the sample and got dissolved into the water during titration or when transferring the liquid.	Keep the flask closed most of the time and use BOD bottles.
Loss of heat during the experiment	Temperature may vary from when it was checked with a thermometer and when the result was been recorded.	Use insulated bottles and close the lid most of the time
Inconsistent temperature of water bath	While taking the flask out from the water bath, its temperature fluctuated which means that the water samples didn't heat up for 300 seconds at a constant temperature.	Use laboratory oven or a hot plate which is a more precise apparatus which can keep the temperature constant.
It was a slow reaction and when titrated it with sodium thiosulfate there was a change in color	Adding sodium thiosulfate in the solution changed the color of the sample sometimes suggesting a slow reaction is taking place.	Add sodium thiosulfate slowly and swirl it after every drop. Let the solution change its color.
Other ions are present in tap water	These ions can influence the reactions taking place in Winkler's method and the result may differ from the actual amount of dissolved oxygen present in water.	We may use distilled water so that the presence of other ions are reduced which affects the result and pump oxygen to the maximum in distilled water.

Further area of scope

Further experiments could be done on other factors like salinity and pH in water to find their effect on the concentration of dissolved oxygen.

Bibliography

Abowei, J. (2010). Salinity, Dissolved Oxygen, pH and Surface Water Temperature Conditions in Nkoro River, Niger Delta, Nigeria. *Advance Journal of Food Science and Technology* , 5.

Catrin Brown, M. F. (2014). *Higher Level Chemistry.* Harlow: Pearson Education Limited.

DETERMINATION OF DISSOLVED OXYGEN BY WINKLER TITRATION. (2006). *12.097 Environmental Chemistry of Boston Harbor – IAP 2006* , 10.

Dickson, A. G. (n.d.). Determination of dissolved oxygen in sea water by Winkler titration. *WHP Operations and Methods — November 1994*, 11.

EPA (United States Environmental Protection Agency. (2012, March). Retrieved from https://archive.epa.gov: https://archive.epa.gov/water/archive/web/html/vms52.html

FISH 503 Advanced Limnology (University of Idaho, Moscow Idaho Campus) . (n.d.). Retrieved from https://www.webpages.uidaho.edu: https://www.webpages.uidaho.edu/fish503al/002%20Oxygen/FISH%20503%20Winkler%20titration%20lab%20day%20I.pdf

Herbold, N. H. (2003). *Dissolved Oxygen.* Retrieved from https://www.sciencedirect.com: https://www.sciencedirect.com/topics/earth-and-planetary-sciences/dissolved-oxygen

Mooney, C. (2016, April). *The Washington Post.* Retrieved from https://www.washingtonpost.com: https://www.washingtonpost.com/news/energy-environment/wp/2016/04/28/global-warming-could-deplete-the-oceans-oxygen-levels-with-severe-consequences/

Popek, E. (2018). *Dissolved oxygen.* Retrieved from https://www.sciencedirect.com: https://www.sciencedirect.com/topics/earth-and-planetary-sciences/dissolved-oxygen

Prior, R. (2019, August Saturday). Retrieved from https://edition.cnn.com: https://edition.cnn.com/2019/08/16/us/alaska-salmon-hot-water-trnd/index.html

3. HOW DOES THE RATE OF FERMENTATION IN YEAST VARY BASED ON THE TEMPERATURE?

Author: Manning Chen

Research Question

The goal of this investigation is to answer the question: How does the rate of fermentation in yeast vary based on the temperature?

Background

Fermentation is an anaerobic biochemical process in which sugars such as glucose or fructose are broken down by yeasts to produce energy for these yeasts. Yeast is not a chemical catalyst for fermentation but instead a living fungus that performs the fermentation. There are many different types of fermentation, however, the one that I am concerned with in this experiment is alcohol fermentation. This type of fermentation is used in the production of a variety of things such as bread or beer. In alcohol fermentation, sugars (usually glucose) are broken up into ethanol alcohol and carbon dioxide (Helmenstine). The net formula for this is presented below:

$$C_6H_{12}O_6\ (aq) \rightarrow 2C_2H_5OH\ (aq) + 2CO_2\ (g)$$

For this investigation, to measure the rate of fermentation, I can use fermentation tubes which traps the CO_2 gas that is produced through the fermentation and measure the amount of CO_2 that is produced in what time. By know the temperature, pressure, and volume of the gas, one can use the ideal gas law in order to find the number of moles of CO_2 gas present.

Temperature is the measure of average kinetic energy, or the average speed of all a sample of molecules. As temperature is increased, the rate of a reaction is also usually increased because an increase in temperature causes increase in the frequency of collision and also increases the energy possessed by each particle making it so that it is more likely to react.

Materials List

- active dry yeast
- distilled water
- 1500 cm^3 volumetric flasks
- scoopula
- funnel
- analytical balance (± 0.001g)
- Anova Precision Cooker (Sous-vide cooker) (± 0.1°C)
- 135.12$\pm$0.01 grams of fructose, $C_6H_{12}O_6$ (0.75 moles)
- Thermometer (± 0.5°C)
- 3 50.0$\pm$0.1 cm^3 graduated cylinders
- 3 fermentation tubes
- large container (for pseudo water bath) (able to comfortably contain three fermentation tubes)
- Online source for atmospheric pressure

Variables

Variables	Variable Type	How is it Manipulated/Measured?
Temperature of fermentation	Independent	The temperature of the fermentation will be maintained using a pseudo water bath using a sous-vide cooker. What a sous-vide cooker does is circulates the water in the water bath and regulates it a precise temperature. To ensure that the temperature of the water remains constant and the sous-vide machine is regulating accurately, a thermometer was used to double-check the temperature of the water.
Moles of CO_2 produced (rate of fermentation)	Dependent Variable	The number of moles of CO_2 will be not be directly measure but instead be calculated by measuring the volume of the CO_2 produced, the pressure, and the temperature and then using a gas law.

Controlled Variable	How will it be controlled?	Justification
Type of Sugar	Fructose was used as the sugar for all the trials	Different sugars are broken down differently yeast and will have different rates of fermentation
Concentration of Sugar Solution	All of the fructose solution were 0.5 moles dm^{-3}	Having solutions of different concentrations will affect access to sugar and rate of fermentation
Fermenting Time	All trials were given 30 minutes to ferment	Longer or shorter fermentation times will lead to more or less carbon dioxide to be produced
H_2O impurities	Distilled water was used to make solution	Impurities within solvent could interfere with process of fermentation and reduce rates
Pressure of CO_2 gas	The atmospheric pressure was recorded independently in the middle of each trial from an online source.	The pressure affects the volume of the CO_2 in the fermentation tube. All the tubes were allowed to ferment in close proximity to each in order to ensure that they are experiencing similar atmospheric pressures. Since there was no way of measuring the pressure of the gas directly, an online source was used to obtained the atmospheric pressure.

Procedure

1. First prepare a 1500 cm³ of 0.5 moles dm⁻³ solution of fructose ($C_6H_{12}O_6$). The directions below explain how to do this:
   a. First mass 0.75 moles (135.12 grams) of fructose.
   b. Place this sample of fructose in a 1500 cm³ volumetric flask using a funnel.
   c. Fill the rest of the volumetric flask up with distilled water.
2. Measure the height of the fermentation tube's mouth when standing upright and fill up the large container with water until the water level reaches just below that height (approximately 1.5 to 2 cm below).
3. Let the water sit for an hour or until its temperature becomes constant with the room's temperature.
4. In each of the three fermentation tubes, place 0.250 grams of yeast.
5. Then measure out 50 cm³ of fructose solution and pour as much possible into the fermentation without overfill and subsequently tilt the fermentation tube back so the fructose solution with the yeast occupies the elevated shaft of the fermentation tube. When doing so, ensure that no air bubbles are trapped in the shaft of the tube.
6. After there are no air bubbles left, set the fermentation tube upright and pour the rest of the 50 cm³ of fructose solution into the fermentation tube.
7. Repeat steps 5 and 6 for the other two fermentation tubes with yeast in them and then set all three fermentation tubes with the fructose-yeast mixture into the water bath for 30 minutes to ferment.
8. Halfway in between, after 15 minutes, use an online source to find and record the local atmospheric pressure of the lab.
9. After half and hour is up, take the fermentation tubes out of the water and then for each tube, pour the contents of the tube into a graduated cylinder and record the volume of the mixture remaining.
10. Using the sous-vide cooker, heat the temperature of the water up to 30° Celsius and using a thermometer, wait for the temperature of the water to become constant.
11. Now repeat steps 4-9 for the 30° Celsius water controlled by the sous-vide cooker.
12. Repeat steps 10-11, by setting the sous-vide cooker to 35.0°, 32.5°, 37.5°, 40.0°, and 45.0° Celsius.

Raw Data

T1: Fermentation (25.4±0.5°C, 30.07±0.01 inHg)		
Volume of 0.5 moles dm⁻³ Fructose and Yeast mixture (±0.5 cm³)		
Trials	Initial Volume	Final Volume
1	50.0	49.5
2	50.0	49.0
3	50.0	49.3

T2: Fermentation (30.0±0.1°C, 30.06±0.01 inHg)

Volume of 0.5 moles dm^{-3} Fructose and Yeast mixture (±0.5 cm^3)		
Trials	Initial Volume	Final Volume
1	50.0	48.8
2	50.0	49.0
3	50.0	48.8

T3: Fermentation (32.5±0.1°C, 29.95±0.01 inHg)

Volume of 0.5 moles dm^{-3} Fructose and Yeast solution (±0.5 cm^3)		
Trials	Initial Volume	Final Volume
1	50.0	47.9
2	50.0	48.5
3	50.0	48.7

T4: Fermentation (35.0±0.1°C, 30.05±0.01 inHg)

Volume of 0.5 moles dm^{-3} Fructose and Yeast solution (±0.5 cm^3)		
Trials	Initial Volume	Final Volume
1	50.0	47.2
2	50.0	47.2
3	50.0	46.9

T5: Fermentation (37.5±0.1°C, 29.94±0.01 inHg)

Volume of 0.5 moles dm^{-3} Fructose and Yeast solution (±0.5 cm^3)		
Trials	Initial Volume	Final Volume
1	50.0	48.0
2	50.0	48.1
3	50.0	47.8

T6: Fermentation (40.0±0.1°C, 30.04±0.01 inHg)

Volume of 0.5 moles dm^{-3} Fructose and Yeast solution (±0.5 cm^3)		
Trials	Initial Volume	Final Volume
1	50.0	49.0
2	50.0	48.7
3	50.0	49.2

T7: Fermentation (45.0±0.1°C, 30.03±0.01 inHg)		
Volume of 0.5 moles dm⁻³ Fructose and Yeast solution (±0.5 cm³)		
Trials	Initial Volume	Final Volume
1	50.0	49.1
2	50.0	49.5
3	50.0	49.0

Qualitative Observation

During the experiment, I noted that the fructose was easily able to dissolve into the solution. During fermentation, it seemed that a majority of the yeast rose to the top of the shaft of the fermentation tube, and foamy bubbles began to form and accumulate there.

Processed Data

To calculate the number of moles of CO_2 produced, the ideal gas law can be used: $PV = nRT$ where P is pressure in kilopascals, V is volume of the gas in cubic decimeters, n is number of moles of gas, T is the temperature in Kelvin, and R is the ideal gas constant (8.314 $J \cdot K^{-1} \cdot mol^{-1}$ or 8.314 $kPa \cdot dm^3 \cdot K^{-1} \cdot mol^{-1}$). This formula can be rewritten so that it is solved for n which is: $n = \dfrac{PV}{RT}$.

To demonstrate this, I will be doing the calculations with trial 1 of the first fermentation which is at room temperature or 25.4°C.

$$V = \text{Initial} - \text{Final Volume} = 50.0 - 49.5 = 0.5 \ (\pm 1.0) \ cm^3 \div 1000 = 5 \times 10^{-4} (\pm 1 \times 10^{-3}) \ dm^3$$

$$P = 30.07 \ (\pm 0.01) \ inHg \times 3.386 \ kPa \cdot inHg^{-1} = 101.83 \ (\pm 0.03) \ kPa$$

$$T = 25.4°C + 273 = 298.4 \ (\pm 0.5) \ K$$

$$n = \frac{PV}{RT} = \frac{101.83 \cdot 5 \times 10^{-4}}{8.314 \cdot 298.4} = 2. \times 10^{-5} \text{ moles of } CO_2$$

Uncertainty calculation

Since the calculation is just multiplication and division, the fractional uncertainties must be added to calculate the absolute uncertainty

$$\text{Fractional uncertainty} = \frac{1 \times 10^{-3}}{5 \times 10^{-4}} + \frac{0.03}{101.83} + \frac{0.5}{298.4} = 2$$

$$\text{Absolute uncertainty} = 2 \times 10^{-5} \cdot 2 = 4 \times 10^{-5}$$

Therefore $n = 2 \times 10^{-5} \pm 4 \times 10^{-5}$ moles of CO_2 were produced for trial 1 for the room temperature fermentation.

Here is the rest of the processed data:

Amount of CO_2 produced (moles)							
Temp.	25.4°C	30.0°C	32.5°C	35.0°C	37.5°C	40.0°C	45.0°C
Trial 1	2×10^{-5}	4.8×10^{-5}	8.4×10^{-5}	1.1×10^{-4}	7.9×10^{-5}	3.9×10^{-5}	3×10^{-5}
Trial 2	4.0×10^{-5}	4.0×10^{-5}	6.0×10^{-5}	1.1×10^{-4}	7.5×10^{-5}	5.1×10^{-5}	1×10^{-5}
Trial 3	2×10^{-5}	4.8×10^{-5}	5.2×10^{-5}	1.2×10^{-4}	8.6×10^{-5}	3.1×10^{-5}	3.8×10^{-5}
Average	3×10^{-5}	4.6×10^{-5}	6.5×10^{-5}	1.2×10^{-4}	8.0×10^{-5}	4.0×10^{-5}	3.1×10^{-5}
Uncert.	$\pm4.0\times10^{-5}$	$\pm4.0\times10^{-5}$	$\pm4.0\times10^{-5}$	$\pm4\times10^{-5}$	$\pm3.9\times10^{-5}$	$\pm3.9\times10^{-5}$	$\pm3.8\times10^{-5}$

Here is the average number of moles of CO_2 produced presented in a graph:

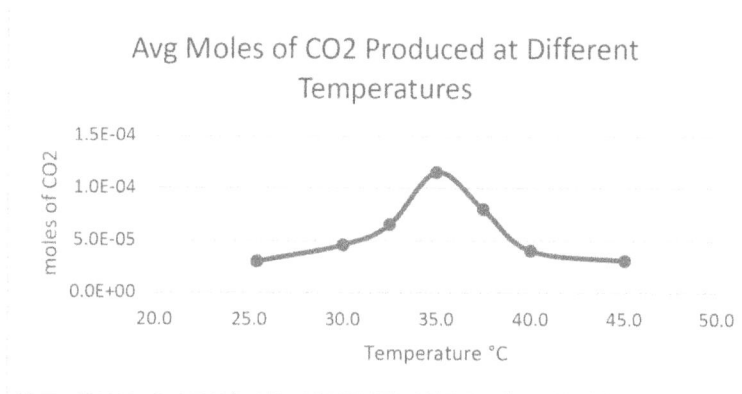

Avg Moles of CO2 Produced at Different Temperatures

Since all trials were allowed to ferment for the same duration of time, this graph can also be considered a representation of the rate of fermentation or how many moles of CO_2 was produced per 30 minutes.

Based on the graph, it seems that the rate of fermentation for fructose increases as temperature increase until it reaches a maximum production rate at around a temperature of 35°C and starts decreasing from there. Although 35.0°C isn't the optimal temperature, it's the closest data value to the maximum production. This means that the ideal temperature for the fermentation of fructose to occur is between 32.5°C and 37.5°C. After the temperature increases past 35°C, the rate of fermentation begins to decrease. Something to note is that the rate of fermentation increases and decreases more drastically when closer to 35°C then when the temperature is far away.

Due to the changing nature of the trend, I will be analyzing the increasing portion of the graph first and the decreasing portion of the graph separately. If only the increasing portion of the graph is focused on, or the first four data points:

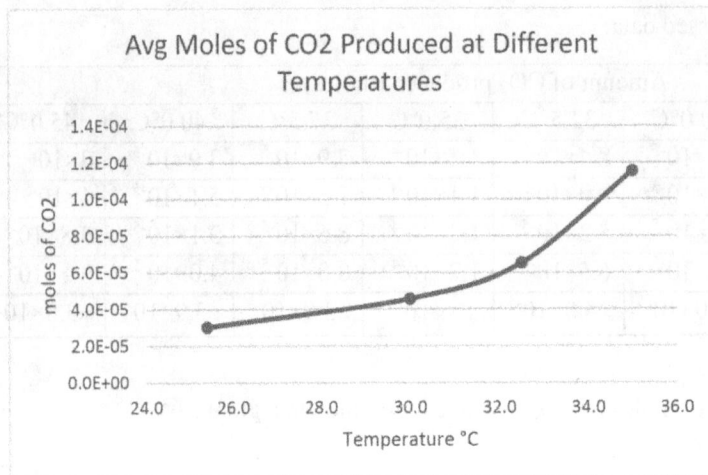

Avg Moles of CO2 Produced at Different Temperatures

In order to linearize the graph, the reciprocal moles of carbon dioxide produced (rate of fermentation) can be graphed against time

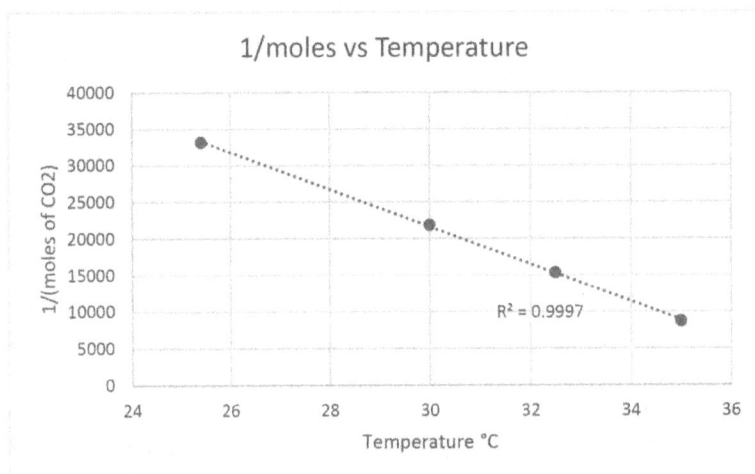

1/moles vs Temperature

$R^2 = 0.9997$

From this graph, it can be seen that there is a strong negative correlation between the average of the reciprocals of the number of moles of CO_2 produced and the temperature of the reaction. With an R^2 value of 0.997, the data points line up virtually in a straight line. However, the problem with this graph is that it is only the graph of the averages and doesn't take into account all of the trials that were done. If we instead graph all the reciprocal of the rate of all the trials versus the temperature, the R^2 value is not as consistent with an R^2 value of 0.779:

1/moles vs Temperature 2

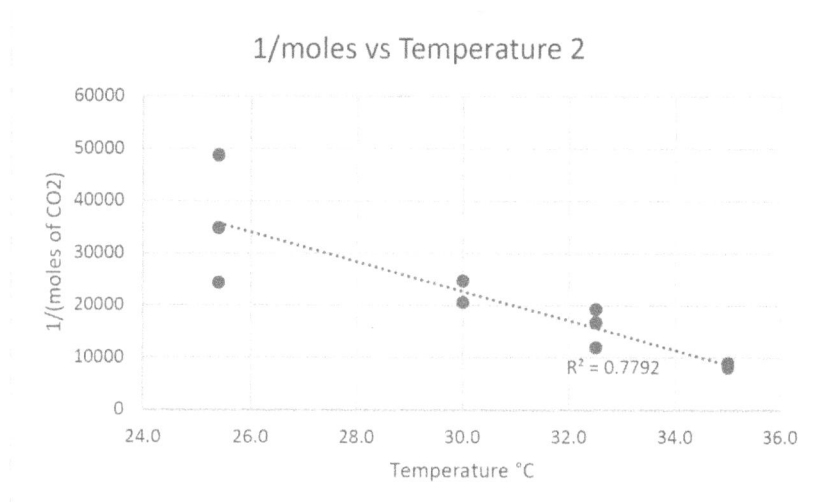

When considering all of the trials, an alteration that linearizes the scatter plot is instead graphing the natural log of the rate of fermentation against the temperature of fermentation:

ln(moles of CO2) vs Temperature

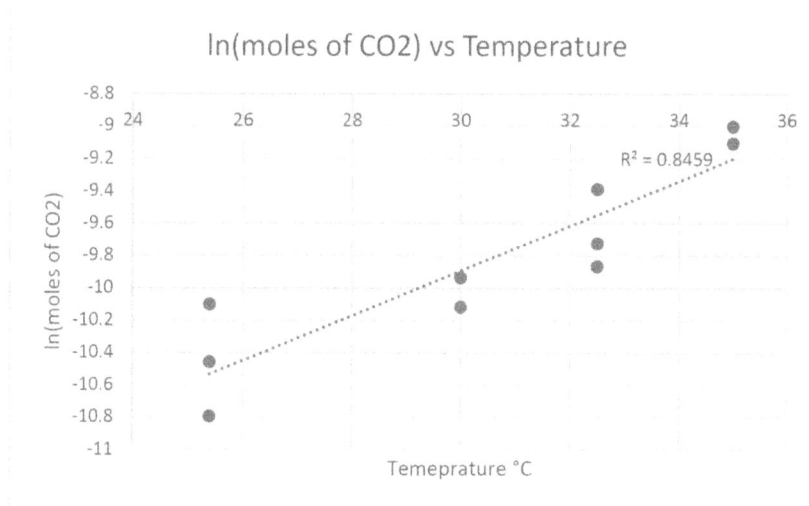

Using this method of linearization, there is a improved R^2 value of 0.8459 giving a relatively strong positive correlation between the natural log of the number of moles of CO_2 produced and the temperature. What this means is that as temperature increases, the rate at which fermentation occurs also increases exponentially. This can be explained through Maxwell-Boltzmann distributions. In the figure below:

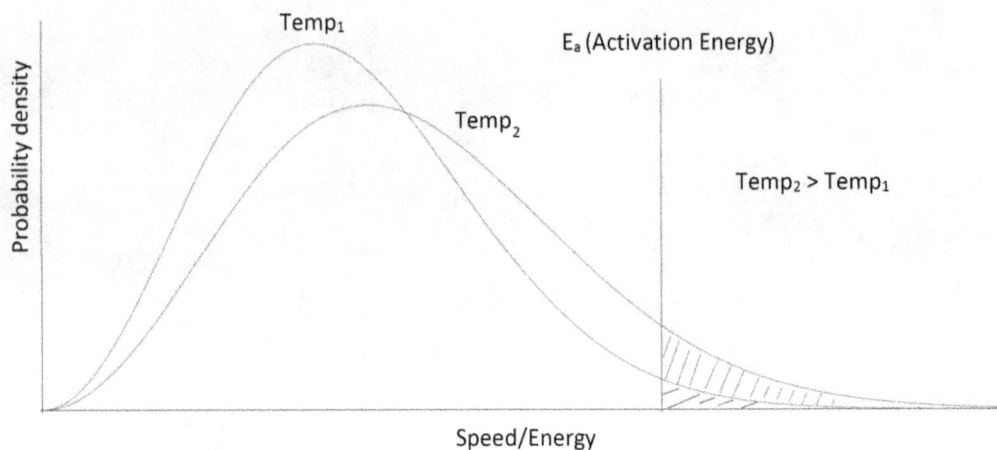

The Maxwell-Boltzmann distribution gives probability distributions of the speed of gases at a specific temperature (Maley). Since temperature is the measure of the average speed or kinetic energy of something, the average speed/Energy for a Maxwell-Boltzmann distribution corresponds with its temperature. What an increase in temperature does is flatten out the distribution, giving it a smaller highest point and more spread meaning that it has a higher probability to have molecules with higher speed and more Energy. In a reaction this is helpful because molecules have a higher average speed meaning that there will be more frequent collisions that could potentially result in a reaction taking place. Also, the increase in temperature causes there to be a higher probability of molecules that have enough energy to actually react if they do collide causing successful reactions to happen more frequently. Because of these two things, an increase in temperature causes more frequent successful reactions allowing for the rate of the reaction to also increase exponentially.

Under normal circumstances, we would expect to see the rate of the reaction to continue to increase exponentially along with temperature, however, that is not the case. The data shows that past 35.0°C the number of moles of CO_2 produced in 30 minutes begins to decreases, creating the maximum production of CO_2 at 35.0°C as can be seen in Graph 1. This is because there is a biological aspect at play: yeast. The decrease in rate of fermentation after the 35.0°C point is caused by the denaturing of the yeast or the shutdown of biological processes within the yeast because the temperature gets too high.

Conclusion

The results of the experiment show that as the temperature at which fermentation occurs increase, the number of moles of carbon dioxide gas produced as well as the rate of fermentation also increase exponentially with it. From a particle standpoint, this is due to the fact that an increase in temperature causes all of the particles to speed up meaning there are more frequent collisions between particles allowing for a greater chance of a successful reaction. The increase in temperature also increases the energy of all the particles meaning that there is a higher probability that if two molecules do collide, they collide with the energy necessary for a reaction to actually occur. Because of these two increases in probability of reaction, the overall rate of the reaction increases as temperature increases; however, the rate of fermentation does not increase indefinitely due to the biology of the yeast. The yeast reaches its maximum rate of fermentation at 35.0°C. Past this temperature, the rate of fermentation begins to decrease due to the shutdown of cellular processes within the yeast at higher temperature. As the temperature continues to increase the yeast becomes more denatured, producing even less CO_2. According to "Dough Fermentation and Temperature," the optimum temperature for fermentation to occur is at 35.0°C which is supported by the results of this experiment since the fast rate of fermentation occurred at 35.0°C. However, in reality, the optimum temperature for fermentation to occur depends on the strand of yeast that was used. Different yeasts, have different temperatures in which their cellular processes began to shut down. Although, it cannot be concluded that the denaturing of this yeast begins at exactly 35.0°C, from the data it can concluded that the temperature at which the yeast begins to stop working is between 32.5°C and 37.5°C.

Evaluation

Strengths

The greatest strength of my experiment my ability to adapt to the lack of water bath even though I still needed some way to maintain a constant temperature. Due to the lack of accessibility to a water bathe, I instead used a sous-vide cooker, which is a cooking method/device that precisely and consistently maintains a constant temperature for water ("What is Sous Vide"). This was used to maintain a consistent temperature through out the duration of the reaction. An actual thermometer was also used in order to ensure that the temperature of the water was kept constant at the intended value.

Weaknesses

Weakness	Effects on results	Possible modification
The pressure that was used for calculation was the atmospheric pressure from an online source.	Because the pressure used for calculation was not the actual pressure of the carbon dioxide gas within the fermentation and instead the atmospheric pressure of the lab, the calculations done with this number will cause any calculated values to be skewed.	Somehow find a way to directly measure the pressure of the gas in the tube using proper lab equipment.
There were no volume markings on the fermentation tube, so instead to directly find the volume that the gas	Since the change of volume of the fructose solution wouldn't directly reflect the volume occupied by the carbon dioxide gas, it would affect any	I could instead use fermentation tubes with volume markings on them, or I could make my own markings on the fermentation tubes myself.

occupied, I took the change in volume of the fructose solution.	calculations done with that value. However, this discrepancy would not affect the overall trend that is observed.	
Fermentation was not able to go to completion	In my investigation, I assumed that 30 minutes fermentation time would be enough time, however, someone of the trials barely fermented any. Also it takes a while for the fructose-yeast solution to reach the temperature of the sous-vide bath, especially for the higher temperatures.	Let every single trial ferment for a longer period of time, maybe an entire hour, instead of just half an hour.

Extensions

A possible extension to my investigation would be to somehow figure out the activation energy necessary for fermentation with fructose to occur. This could be done by writing a rate equation through experimental testing for fermentation and then creating an Arrhenius plot, by doing the fermentation at different temperatures like what was done in this investigation but also measuring the final concentration of the fructose solution in order to get the rate of the reaction in moles·dm^{-3}·sec^{-1} for each temperature. And then an Arrhenius plot can be made and the activation energy could be extracted from the slope of its trendline.

Works Cited

"Dough Fermentation and Temperature." *The Artisan*, 25 Dec. 2001, www.theartisan.net/dough_fermentation_and_temperature.htm

Helmenstine, Anne Marie. "What Is Fermentation? ." *ThoughtCo*, ThoughtCo, 22 Jan. 2019, www.thoughtco.com/what-is-fermentation-608199.

Maley, Adam. "Maxwell-Boltzmann Distributions." *Chemistry LibreTexts*, Libretexts, 5 June 2019, chem.libretexts.org/Bookshelves/Physical_and_Theoretical_Chemistry_Textbook_Maps/S upplemental_Modules_(Physical_and_Theoretical_Chemistry)/Kinetics/Rate_Laws/Gas_P hase_Kinetics/Maxwell-Boltzmann_Distributions.

"What Is Sous Vide?" *Anova Culinary*, Anova , anovaculinary.com/what-is-sous-vide/.

4. THE EFFECT OF ELECTROLYTE CONCENTRATION ON THE CELL POTENTIAL OF A GALVANIC CELL

Author: Simone Minervini
Moderated Mark: 22/24

Personal motivation:

Batteries are a big part of our lives: from classic batteries that power TV remotes to thin lithium batteries present in smartphones. As a result, the world of batteries is under constant pressure to provide lighter, cheaper and longer-lasting products. As an aspiring material engineer, I have this field strongly at heart. Besides, who doesn't hate the incredibly short lifespan of our laptops and smartphones? Therefore, I decided to focus my investigation on batteries.

In chemistry, batteries are referred to as "galvanic" or "voltaic" cells (electrochemical cells which generate electricity from a redox chemical reaction). [8] They consist of two separate half-cells, in which two different metal electrodes are immersed in their own salt solution (electrolytes) connected by a "conduction" wire and a "salt bridge". As a spontaneous redox reaction occurs, electrons flow in the external circuit (conduction wire) from one half-cell to the other, producing a potential difference referred to as "cell potential". [9] To keep electrical neutrality, the salt bridge allows ions to flow from one half-cell to the other (balances charge caused by electron flow). [10]

Investigating what aspect of galvanic cells I wanted to test, I discovered that cell potential can be measured outside standard conditions (298°K, 100KPa and 1 molar concentration solutions) and was eager to try. I therefore decided to investigate the effect of electrolyte concentration on cell potential (E), answering the question: How does decreasing the concentration of the reduced electrolyte ($CuSO_4$) affect the cell potential of galvanic cells, with $AlCl_3$ and $ZnSO_4$ half-cells?

Two galvanic cells are tested (using 2 different anodes to have a broader understanding):

1. $Cu_{(s)}$ in $CuSO_4$ (aq.) connected with $Al_{(s)}$ in $AlCl_3$ (aq.)   [written as: $Zn_{(s)}$ / Zn^{2+}(aq.) II Cu^{2+}(aq.) / $Cu_{(s)}$]

2. $Cu_{(s)}$ in $CuSO_4$ (aq.) connected with $Zn_{(s)}$ in $ZnSO_4$ (aq.)   [written as: $Al_{(s)}$ / Al^{3+}(aq.) II Cu^{2+}(aq.) / $Cu_{(s)}$]

The net ionic equations for the two different galvanic cells are:

1. $3Cu_{(aq.)}^{2+} + 2Al_{(s)} \rightarrow 2Al_{(aq.)}^{3+} + 3Cu_{(s)}$ Where 6 moles of electrons are transferred

---

[8] Geoffrey Neuss, *Chemistry for the IB diploma* (Oxford: Oxford University Press, 2014), 73.
[9] LibreTexts, "The Cell Potential," *Chemistry LibreTexts*, December 10, 2016, Accessed November 25, 2017, https://chem.libretexts.org/Core/Analytical_Chemistry/Electrochemistry/Voltaic_Cells/The_Cell_Potential.
[10] Geoffrey Neuss, *Chemistry for the IB diploma* (Oxford: Oxford University Press, 2014), 73.

2. $Cu^{2+}_{(aq.)} + Zn_{(s)} \rightarrow Zn^{2+}_{(aq.)} + Cu_{(s)}$ <u>Where 2 moles of electrons are transferred</u>

The net ionic equations show that, in the galvanic cells, Al and Zn oxidize (lose electrons). This occurs at the anode (the negative electrode) where the metal electrodes dissolve, become aqueous ions (Al^{3+} and Zn^{2+}). Meanwhile, in both galvanic cells, the aqueous Cu^{2+} ions present in the electrolyte solution reduce (gain electrons). This occurs at the cathode (the positive electrode) where the Cu^{2+} ions become Cu metal, depositing on the electrode. Cu^{2+} reduces whilst Al and Zn oxidize because of their standard electrode potential (E_0): Al^{3+} = -1.66V, Zn^{2+} = -0.76V and Cu^{2+} = 0.34V. [11] As this value indicates the reduction potential, Cu^{2+} is more prone to reduction. "Nernst Equation": $E = E_0 - \frac{RT}{nF} \ln \left(\frac{[oxidized\ species]^y}{[reduced\ species]^x} \right)$ is used to calculate the non-standard cell potential (E). [12] Where, E_0 = standard electrode potential, R = gas constant (8.31 J/K/mol), T = temperature in °K, n = moles of transferred electrons in the reaction, F = Faraday's constant (96500 C/mol), y and x = the respective coefficients of the oxidized and reduced species].

Variables:

Independent Variable: $CuSO_4$ of concentration

Dependent Variable: Cell potential of the galvanic cell = E_{cell} measured in volts (V)

Controlled variables:

Variable	Effect if variable is not controlled	How was it controlled?
$AlCl_3$ and $ZnSO_4$ concentration	Alters the cell potential produced	Use 1.00M concentration (assumed to remain constant)
Temperature	Alters the cell potential produced (higher temperature lowers E_{cell})	Measure temperature before each trial, recording any change.
Voltmeter resistance	Alters the *measured* cell potential	Use the same voltmeter
Salt bridge	Could alter ion flow (altering cell potential)	Use the same 3M KNO_3 salt bridge
Conduction wires resistance	Alters the *measured* cell potential	Use the same conduction wires
Crocodile clips resistance	Alters *the measured* cell potential	Use the same crocodile clips
Time of reaction	Cell potential fluctuates, meaning imprecision	Measure E_{cell} after 30 seconds
Electrodes oxidization	Impurities affect the cell potential	Polish, rinse and dry electrodes

---

[11] Geoffrey Neuss, *Chemistry for the IB diploma* (Oxford: Oxford University Press, 2014), 75.
[12] Geoffrey Neuss, *Chemistry for the IB diploma* (Oxford: Oxford University Press, 2014), 149.

Volume of electrolyte		Measure 20cm³ of solution
Electrode surface area	Theoretically affectless on cell potential (yet controlled to have homogeneous conditions)	Use electrodes 2 x 6 x 0.3cm
Depth of electrode immersion in electrolyte		Immerge the electrolyte by 3cm

Apparatus:

2.1 - Material required to prepare AlCl₃ solution:

- Digital scale (for digital apparatus: least count = uncertainty = 0.001g)
- Stirring rod
- Spatula
- 50cm³ beaker (measurements are not taken with this apparatus)
- 50cm³ graduated cylinder (least count = 1cm³) (±0.5cm³)
- 50cm³ burette (least count = 0.1cm³) (±0.05cm³)

2.2 - Material required to dilute the given 1.00M CuSO₄ solution:

- 50cm³ beaker (measurements are not taken with this apparatus)
- 50cm³ burette (least count = 1cm³) (±0.5cm³)
- Four 50cm³ graduated cylinders (least count = 0.1cm³) (±0.05cm³)
- Stirring rod

2.3 - Material required to measure cell potential:

Voltmeter (for digital apparatus: least count = uncertainty = ±0.01V)

- 2 conduction wires + 2 crocodile clips
- Stopwatch
- KNO_3 salt bridge of 3M concentration
- Two 30cm³ beakers (measurements are not taken with this apparatus)
- Sand paper
- Aluminum electrode (aluminum foil)
- Copper metal electrode
- Zinc metal electrode
- 2 thermometers (least count = 0.1°C) (±0.05°C)

Chemicals used: 1.00M copper (II) sulfate (then diluted to make the different concentrations tested), 1.00M aluminum chloride and 1.00M zinc sulfate. Both copper (II) sulfate and zinc sulfate were given by the lab technician, quoting 3 significant figures. Instead, the aluminum chloride solution was prepared dissolving solid powder in distilled water.

$CuSO_4$, $AlCl_3$ and $ZnSO_4$ were chosen because of their low hazard and availability.

Safety and ethics: All waste electrolytes are disposed in the appropriate container for hazardous chemicals which is then processed and safely disposed. For the experimental procedure, safety goggles and lab coats are worn at all times to avoid possible skin or eye contact with the chemicals. It is ensured that the electric components do not come into contact with any liquid. Low voltage/current produced and low toxicity of the chemicals ensure a safe procedure.

Procedure and apparatus set-up:

Preparing 30cm³ of 1.00M aluminum chloride solution:

1. Take the necessary safety measures and prepare the clean equipment (apparatus 2.1).
2. To calculate the necessary mass of $AlCl_3 \bullet 6H_2O$ to prepare 30cm³ of solution the following calculation is carried out: *Molar mass × volume (in dm^3)*. Hence, since the molar mass of hexahydrate compound is 241 g/dm³, the mass required is $241 \times \frac{30}{1000} = 7.23g$.
3. Place the empty 50cm³ beaker on the scale and record the mass. Add 7.23g of $AlCl_3 \bullet 6H_2O$ and record the new mass. Add 20.0cm³ of distilled water using a burette and stir with a stirring rod until the powder is fully dissolved. Attention, *this is an exothermic process.*
4. *Allow the solution to cool before proceeding,* then transfer the solution to a 50cm³ graduated cylinder. Fill, with distilled water, the required volume to reach the 30cm³ mark.

5. Pour 20cm^3 of the solution into a clean 30cm^3 beaker. This is ready to be tested.

Volume of original 1.00M CuSO$_4$ solution (cm^3)	Volume of distilled water added (cm^3)	Concentration of CuSO$_4$ solution produced (M)
15.0	15.0	0.50
9.0	21.0	0.30
3.0	27.0	0.10
1.5	28.5	0.05

Note: The significant figures quoted in the concentration of the solution respect the lowest number of significant figures dealt with in the preparation.

Diluting 1.00M CuSO$_4$ to obtain 30cm^3 of different CuSO$_4$ concentrations

1. Take the necessary safety measures and prepare the clean equipment (apparatus 2.2).

2. Use $C_1V_1 = C_2V_2$ (left-hand side represents the original concentration and volume while the right-hand side represents the new concentration and volume) to find the volumes needed to obtain the chosen concentrations (see table in step 3 below). The concentrations were chosen to ensure measurable changes in cell potential.

3. The volumes calculated are reported in the table below:

4. The reported volumes of "1.00M CuSO$_4$ solution" are poured from a burette into 4 different graduated cylinders. The respective volumes of distilled water are added. 30cm^3 of all diluted CUSO$_4$ solutions are produced, ready to be tested in the galvanic cells.

Apparatus set-up of the galvanic cell:

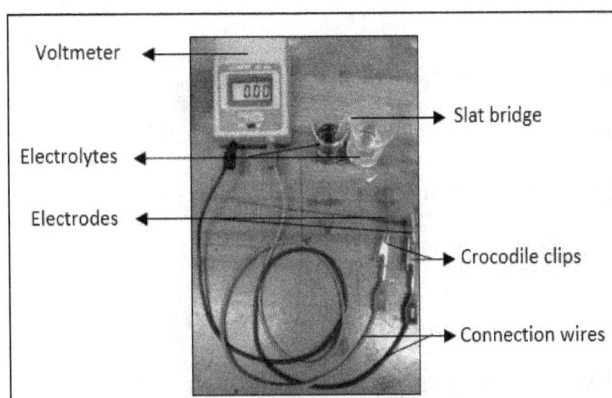

Fig.1. Experimental set up of galvanic cells

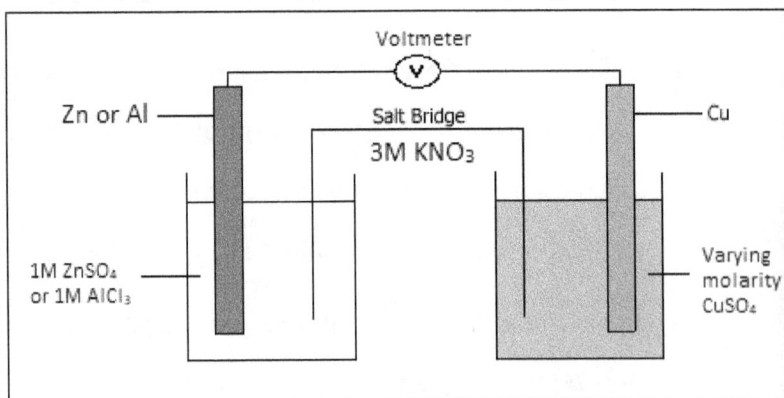

Fig.2. Diagram of galvanic cells

Testing the galvanic cell (procedure):

1. Connect the voltmeter with two conduction wires, each having a crocodile clip at its extremity. Make sure to use the same throughout the whole investigation.

2. Polish the Cu and Zn electrodes to remove any possible oxide layer. Rinse them in distilled water, let them dry and weigh them. Fold the aluminum foil into a small 2 x 6 x 0.3cm rectangle and weigh it. Due to the very low thickness of the foil, the electrode cannot be polished to remove any possible Al_2O_3 layer present.

3. Two galvanic cells will be made: one with Cu and Al with respective solutions and one with Cu and Zn and respective solutions. The following steps represent the cell with Cu and Al.

4. Clip the Cu and Al electrodes with the crocodile clips. The Al (anode) should be clipped to the wire connecting to the negative terminal on the voltmeter and the Cu cathode to the positive terminal.

5. Place the beaker containing 1.00M $AlCl_3$ and the one containing 1.00M $CuSO_4$ next to each other. Connect them with the 3M KNO_3 salt bridge.

6. Insert 3 cm deep the electrodes in their respective electrolytes (Cu in $CuSO_4$ and Al in $AlCl_3$). Start the stopwatch and read the cell potential after 30 seconds.

7. Remove the electrodes, rinse in distilled water, dry and repeat for 4 more trials.

8. After 5 trials change the beaker containing 1.00M $CuSO_4$ to one with a different concentration (0.50M, 0.30M, 0.1.00M and 0.05M) and repeat 5 trails for each.

9. Repeat steps 4 to 8 for the Cu and Zn galvanic cell with Zn electrode and $ZnSO_4$.

10. Clean up the material and dispose of the waste in an appropriate container which will be processed and safely disposed.

Note: Even if 30cm³ of each solution were prepared, only 20 cm³ were tested. This is to avoid preparing the exact amount needed to test and losing some when transferring from one container to another, eventually testing less and not controlling this variable.

Data collection:

Raw Data

Table 1: Cell potential of galvanic cell with **Cu and Al changing the CuSO₄ concentration**

CuSO₄ concentration / M (uncertainty referred in processed data table)	Cell potential, E_{cell} / V (±0.01 V)				
	Trail 1	Trial 2	Trial 3	Trail 4	Trial 5
1.00	1.76	1.77	1.78	1.77	1.77
0.50	1.76	1.75	1.77	1.76	1.76
0.30	1.75	1.75	1.74	1.76	1.75
0.10	1.74	1.74	1.74	1.73	1.75
0.05	1.73	1.73	1.73	1.74	1.72

AlCl₃ concentration was constant (1.00M)   Experimental measured temperature = 298K

Table 2: Cell potential of galvanic cell with **Cu and Zn changing the CuSO₄ concentration**

CuSO₄ concentration / M (uncertainty referred in processed data table)	Cell potential, E_{cell} / V (±0.01 V)				
	Trail 1	Trial 2	Trial 3	Trail 4	Trial 5
1.00	0.99	0.99	1.00	0.99	0.99
0.50	0.99	0.97	0.98	0.98	0.98
0.30	0.97	0.97	0.97	0.97	0.97
0.10	0.96	0.97	0.96	0.98	0.94
0.05	0.96	0.95	0.95	0.95	0.94

ZnSO₄ concentration was constant (1.00M)   Experimental measured temperature = 298K

Processed Data:

The calculation of the CuSO₄ concentration relative uncertainty is reported after the processed data.

Table 3: Processed data table with **average and theoretical** cell potential of Cu and Al galvanic cell

CuSO₄ concentration / M	Natural log of CuSO₄ concentration	CuSO₄ concentration relative uncertainty (%)	Average cell potential, E_{cell} / V (±0.01 V)	*Theoretical cell potential, E_{cell} / V*
1.00	0.00	0	1.77	2.00
0.50	-0.69	2	1.76	1.99
0.30	-1.20	2	1.75	1.98
0.10	-2.30	3	1.74	1.97
0.05	-3.00	5	1.73	1.96

Table 4: Processed data table with **average and theoretical** cell potential of Cu and Zn galvanic cell

CuSO₄ concentration / M	Natural log of CuSO₄ concentration	CuSO₄ concentration Relative uncertainty (%)	Average cell potential, E_{cell} / V (±0.01 V)	*Theoretical cell potential, E_{cell} / V*

1.00	0.00	0	0.99	1.10
0.50	-0.69	2	0.98	1.09
0.30	-1.20	2	0.97	1.08
0.10	-2.30	3	0.96	1.07
0.05	-3.00	5	0.95	1.06

Note:

- **CuSO₄ concentration relative uncertainty (%)** is the same for both the concentration and the natural log of the concentration.

- **Table 3:** the percentage difference between the average cell potential and theoretical cell potential is 12% for all CuSO₄ concentrations (see "calculations").

- **Table 4:** the percentage difference between the average cell potential and theoretical cell potential is 10% for all CuSO₄ concentrations (see "calculations").

Observations:

When inserting the electrodes in the electrolyte solution, no physical change or reaction was observed. Evidently, if the reaction was allowed to carry on for enough time the anode would completely dissolve while the CuSO₄ solution would decolorize (Cu^{2+} ions deposit on the cathode).

Calculations:

Average cell potential, $E_{cell} = \Sigma$ of trials / 5

Ex. for 1.00M CuSO₄, table 4: $\dfrac{0.99+0.99+1.00+0.99+0.99}{5} = 0.99V$

Natural log of CuSO₄ concentration = $ln($CuSO₄ concentration$)$

Ex. for 0.50M CuSO₄, table 4: $ln(0.50) = -0.69$

Percentage difference = $\dfrac{Theoretical\ cell\ potential - average\ cell\ potential}{theoretical\ cell\ potential} \times 100$

Ex. for 1.00M CuSO₄, table 3: $\dfrac{2.00-1.77}{2.00} \times 100 = 12\%$

Theoretical E_{cell} = $E_0 - \dfrac{RT}{nF} \ln\left(\dfrac{[oxidized\ species]^y}{[reduced\ species]^x}\right)$

Ex. for 1.00M CuSO₄ and 1.00M Al₂Cl₃: $2.00 - \dfrac{8.31*298}{6*96500} \ln\left(\dfrac{[1.00]^2}{[1.00]^3}\right) \to 2.00 - 0 = 2.00V$

Uncertainties: The 1.00M concentration of CuSO₄ and ZnSO₄ is assumed to be certain as it was given as a 1.00M solution, without providing uncertainties.

Calculating the relative uncertainty of the CuSO₄ concentration:

<u>Moles relative uncertainty:</u> $\frac{Burette\ uncertainty}{Volume\ of\ 1.00M\ CuSO_4 solution\ added} \times 100$. For example, for the

production of 0.50M CuSO$_4$ solution: $\frac{0.05}{15} \times 100 = 0.3\%$ Hence, the uncertainty is: 0.3% for 0.50M,

0.6% for 0.30M, 1.7% for 0.10M and 3.3% for 0.05M.

<u>Volume relative uncertainty:</u>

$Total\ absolute\ uncertainty = Absolute\ uncertainty\ of\ Burette\ \&\ Graduated\ cylynder$

That is: $0.05 + 0.5 = \pm0.55\ cm^3\ uncertainty.$

$Volume\ relative\ uncertainty = \frac{Total\ absolute\ uncertainty}{Volume\ measured} \times 100.$ Therefore: $\frac{0.55}{30} \times 100 = \pm1.7\%$

Since the "volume measured" is always 30 cm^3 the "volume relative uncertainty" is fixed at ± 1.7%

The FINAL RELATIVE UNCERTAINTY is reported, to 1 significant figure, in tables 3 and 4.

That is the sum of the "moles relative uncertainty" for the corresponding CuSO$_4$ concentration and the "volume relative uncertainty".

Graphical representation:

With the data shown in the processed data tables (tables 3 and 4) it is possible to plot cell potential (E$_{cell}$) against the natural log of the copper sulfate concentration. The natural log is taken to linearize the plot, as the data should follow the equation $E_{cell} = E_0 - \frac{RT}{nF}\ln\left(\frac{[oxidized\ species]^y}{[reduced\ species]^x}\right)$. By expanding the natural log: $E_{cell} = E_0 - \frac{RT}{nF}[\ln([oxidized\ species]^y) - \ln([reduced\ species]^x)]$. The "oxidized species" are the Al^{3+} and Zn^{2+} ions of 1.00M concentration (assumed constant). Hence: $E_{cell} = E_0 + \frac{RT}{nF}x\ln([reduced\ species])$ where E_0 is the y-intercept and $\frac{RT}{nF}x$ the slope.

The slope $\frac{RT}{nF}x$ is composed of constants (R, T, F and the ratio $\frac{x}{n}$), therefore theoretically being identical for both galvanic cells. The ratio $\frac{x}{n}$ is constant because, referring to the net ionic equations in the introduction, it can be appreciated that for the Cu and Al cell, x = 2 and n = 6 whilst for the Cu and Zn cell, x = 1 and n = 2 (where x = the coefficient in front of Cu^{2+} and n = moles of electrons transferred between the two half-cells). Therefore, $\frac{x}{n} = \frac{1}{3}$ for for both galvanic cells.

The processed data allows two separate graphs (one for the Cu and Al galvanic cell and one for the Cu and Zn galvanic cell) to be plotted. They can then be compared to each other and to the respective literature values. *This is sufficient to answer the original research question.*

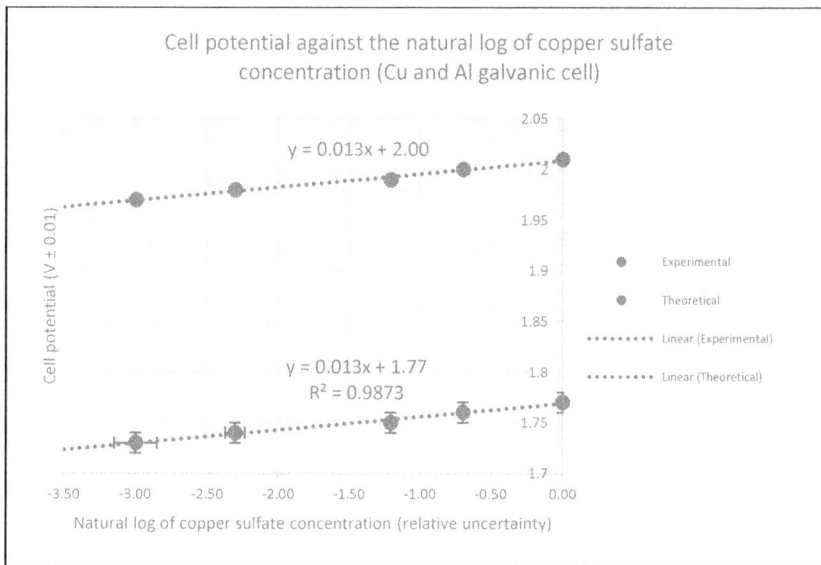

Cell potential against the natural log of copper sulfate concentration (Cu and Al galvanic cell)

y = 0.013x + 2.00

y = 0.013x + 1.77
$R^2 = 0.9873$

Experimental

Theoretical

Linear (Experimental)

Linear (Theoretical)

Cell potential (V ± 0.01)

Natural log of copper sulfate concentration (relative uncertainty)

Fig. 3. Graphical representation of the data for the *Cu and Al galvanic cell*

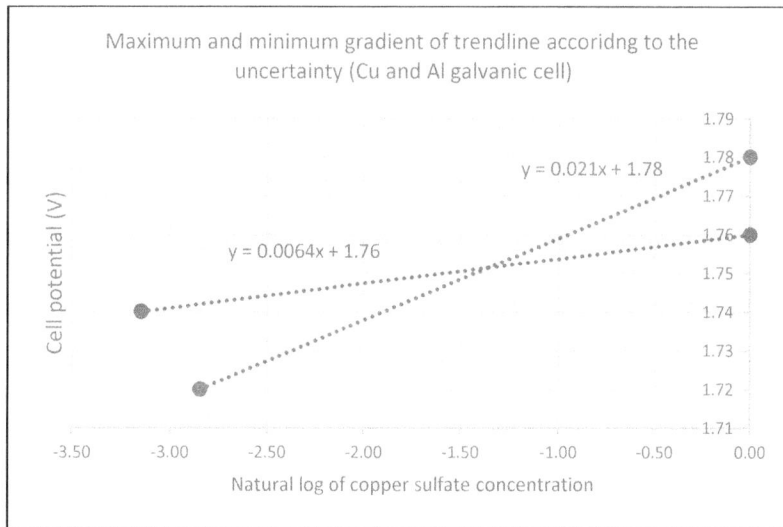

Maximum and minimum gradient of trendline accoridng to the uncertainty (Cu and Al galvanic cell)

y = 0.021x + 1.78

y = 0.0064x + 1.76

Cell potential (V)

Natural log of copper sulfate concentration

Fig. 4. Maximum and minimum gradient of trendline

using the uncertainty (Cu & Al galvanic cell)

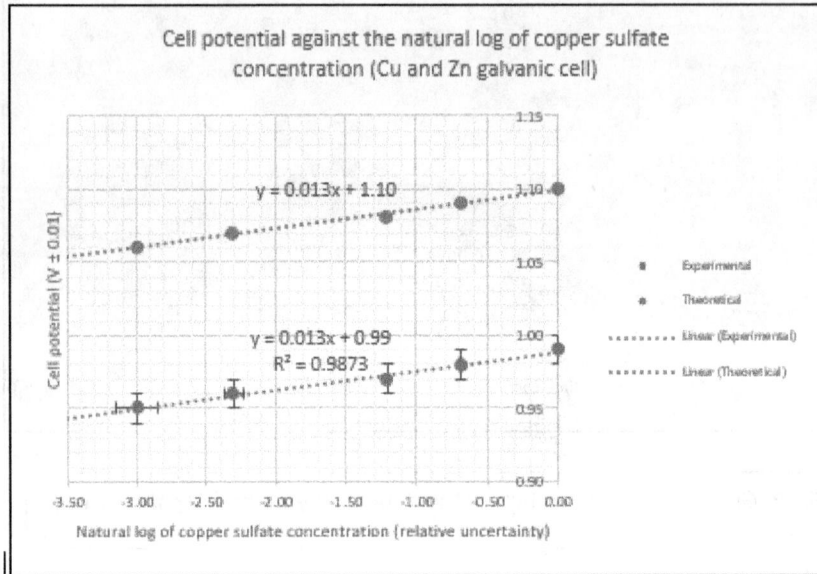

Fig. 5. Graphical representation of the data for the *Cu and Zn galvanic cell*

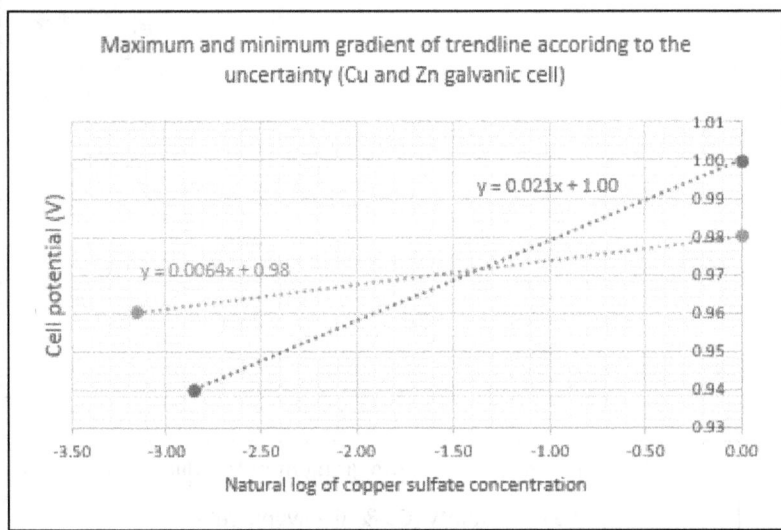

Fig. 6. Maximum and minimum gradient of trendline using the uncertainty (Cu & Zn galvanic cell)

Conclusion and Evaluation:

The analysis will be split in three sections: first analyzing the data from the Cu and Al galvanic cell, then the data from the Cu and Zn galvanic cell and finally comparing the two galvanic cells. This is ultimately followed by the weaknesses and errors present in the procedure.

Cu and Al galvanic cell:

In figure 3, the <u>experimental</u> plot is linear, following the equation: $E = E_0 + \frac{RT}{nF}x \ln([reduced\ species])$ where E_0 is the y-intercept and $\frac{RT}{nF}x$ is the slope of the graph. The <u>experimental</u> E_0 was 1.77±0.01V and the <u>experimental</u> slope was 0.013 (ranging from 0.021 to 0.0064 with max/min slopes, fig 4).

Similarly, the <u>theoretical</u> plot is also linear. The <u>theoretical</u> value of E_0 for this cell was 2.00V, 12% larger than the <u>experimental</u> E_0. This is significantly more than the E$_{cell}$ relative uncertainty $(\frac{E_{cell}\ absolute\ uncertainty}{Measured\ E_{cell}} \times 100 \rightarrow \frac{0.01}{1.77} \times 100 = 0.56\%)$. Despite the possible range given by max/min slopes (fig. 4) the <u>theoretical</u> slope is identical to the <u>experimental</u> slope (0.013).

Cu and Zn galvanic cell:

Figure 5 shows that the <u>experimental</u> plot of the second galvanic cell is also linear where the <u>experimental</u> y-intercept (E_0) is 0.99±0.01V and the <u>experimental</u> slope is 0.013. For this cell, the <u>theoretical</u> value of the E_0 was 1.10V, 10% larger than the <u>experimental</u> value (0.99). This is, again, more than accounted by the ±1.0% E$_{cell}$ relative uncertainty $\left(\frac{0.01}{0.99} \times 100 = 1.0\% \right)$. Finally, as expected, both plots (figures 3 and 5) have the same slope (0.013) matching the <u>theoretical</u> slope. It can therefore be concluded that the slopes portray very high accuracy (matching the <u>theoretical</u> slope) as well as precision (since the two galvanic cells' plots have identical slopes).

Comparing the two galvanic cells

It can be concluded that the <u>experimental</u> plots (figures 3 and 5) portrayed the correct slope yet being vertically shifted downwards (where E$_{cell}$ <u>theoretical</u> > E$_{cell}$ <u>experimental</u>). In this case, the data was precise (as trials were maximum 0.02V apart) but lacked accuracy (since E$_{cell}$ <u>theoretical</u> was 12% and 10% larger than E$_{cell}$ <u>experimental</u> for the respective galvanic cells).

To understand the reason beneath this discrepancy, which is not accounted for by the very low E$_{cell}$ relative uncertainty (±0.56% and ±1.0%), errors and weaknesses must be taken into consideration. Firstly, it can be observed that *systematical errors* were present, explaining why E$_{cell}$ <u>theoretical</u> > E$_{cell}$ <u>experimental</u>. These *systematical errors* are:

1. Infinite resistance assumption: The principal reason beneath the graphical shift present is the resistance of the voltmeter, which is likely not infinite. This *systematical error* lowers the recorded E_{cell}. Using a sophisticated digital voltmeter should reduce this error.

2. Concentration dilution: Using the same electrolyte solution for all trials and for both galvanic cells propagates possible errors made when diluting. This would result in a shift when plotting E_{cell} against ln(concentration). To avoid this, preparing (from dilution) multiple samples of the tested concentrations would ensure that any error in diluting is spotted when seeing a discrepancy in the results (the solution would be re-prepared).

Even though the precision of the data collected was good, there were some *random errors:*

1. Low voltmeter precision: Having a voltmeter which only shows two decimal points limits the precision of the voltage recorded. Since the effect on the voltage value recorded depends on the voltage value itself (affecting the way the voltmeter expresses voltage to two decimal point) this is a *random error.* It could be eliminated using a digital voltmeter which shows voltage to more than two decimal points.

2. Constant concentration assumption: Throughout the investigation, it is assumed that that the concentration of electrolytes is constant throughout the trials. This is not completely true as, through Cu^{2+} deposition, the concentration of $CuSO_4$ decreases as the reaction proceeds (especially damaging at low concentrations like 0.05M). To avoid this, using a new solution for each trial will remove this error (ensuring higher precision).

To extend this investigation one could seek to understand the effect of temperature or metal type on cell potential. It would also be interesting to understand how one could increase a galvanic cell's current, making longer lasting batteries.

Finally, having analyzed the data collected, the answer to the research question is: reducing the concentration of the electrolyte undergoing reduction ($Cu^{2+}_{(aq.)}$) lowers cell potential with a natural logarithmic (ln) relationship.

Bibliography:

LibreTexts. "The Cell Potential." *Chemistry LibreTexts*. December 10, 2016. Accessed November 25, 2017. https://chem.libretexts.org/Core/Analytical_Chemistry/Electrochemistry/Voltaic_Cells/The_Cell_P otential.

Neuss, Geoffrey. *Chemistry for the IB diploma*. Oxford: Oxford University Press, 2014.

5. INVESTIGATING THE EFFECT OF COMMON CHEMICAL COMPOUNDS – ACIDS AND ALKALIS ON THE TENSILE STRENGTH OF CLASSIC PLANT AND ANIMAL FIBRES

Author: Kesavan K
Moderated Mark: 22/24

1. **Aim**

   The aim of this experiment is to investigate the effect of common chemical compounds – acids and alkalis on the tensile strength of classic plant and animal fibres.

2. **Introduction**

   In our common daily life as IBDP students, we go to the laboratory for various reasons. Be it for the IA, or the regularly prescribed experiments, science laboratories are the most frequented places in the school, second only to the playground. In the Chemistry laboratory, we all work with many chemicals, the most used being acids, and alkalis. And we surely wear protective gear made of various kinds of materials. Is a lab coat a prerequisite for a laboratory session? I always refused to wear a lab coat, since it made me feel uncomfortable. We all know is that the chemical compounds we use in the laboratory in some way or the other are harmful to our body, beginning with the skin. To affect the body, the compounds have to first get through the clothes worn, and this depends on the type of fibre the cloth is made from. This thought made me inquisitive and compelled me to investigate the effect of common chemical compounds – acids (Nitric acid) and alkalis (Sodium hydroxide), on certain fibres that are commonly used to manufacture clothes and aprons.

3. **The Research Question**

   Common threads such as cotton, silk and wool are nothing but woven fibres. These fibres have many innate properties, such as fibre morphology, specific gravity, elongation, elastic recovery, tensile strength, electrical conductivity, chemical reactivity, resistance and flammability. This paper focuses on investigating the variation of the tensile strength of a fibre when reacted with acid – nitric acid and alkali – sodium hydroxide of various concentrations. The research question around which the paper revolves is –

   "In what way is the tensile strength of typical plant and animal fibres of controlled lengths affected by timed reactions with NaOH and HNO₃ of various concentrations?"

4. **Background**

   A fibre is essentially a natural or synthetic substance, that is "significantly longer than it is wide." Fibres are widely used in the manufacture of other materials. From the strongest of engineering materials to the clothes we wear daily, everything is made of various kinds of fibres. Synthetic fibres can often be produced very cheaply and in large amounts compared to natural fibres, but for clothing natural fibres can give some benefits, such as comfort, over their synthetic counterparts. Natural fibres include those produced by plants, animals and geological processes. They can be further classified into various categories according to their origin- Vegetable fibres, wood fibres, animal fibres, mineral fibres, biological fibres – fibrous proteins or protein filament. The fibres used in this investigation were cotton, silk and wool.

Cotton is a plant fibre, further categorized as a vegetable fibre. The fibre is made of Lignocellulose – almost pure cellulose. The reaction between an acid and cellulose is called acid hydrolysis of cellulose. During this reaction, the constituents of Lignocellulose – "Lignin and Hemicellulose is broken down into simple fermentable sugars"[13]

Figure 1: Hydrolysis of cellulose

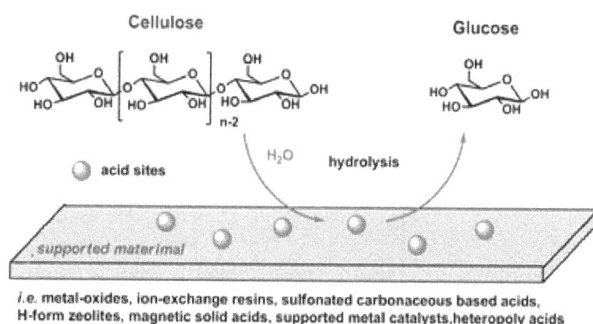

Source: Huang, Yao Bing. "Hydrolysis of cellulose to glucose by solid acid catalysts"[14]

Whereas, when cellulose is exposed to alkalis, it does not react directly. Lignin and the other polysaccharides are stable in alkalis. Since cellulose is a very big molecule with long chains, the minor fragmentation of other components is negligible.[15]

Wool and silk are both animal fibres. Wool may be affected by concentrated acids but does not have any significant effect when exposed to weak acids or diluted mineral acids. The chemical nature of wool keratin is such that it is particularly sensitive to alkaline substances.

---

[13] Rosa, Silvia Morales-dela, "Complete chemical hydrolysis of cellulose into fermentable sugars via ionic   liquids and antisolvent pretreatments", *ChemSusChem*, DOI: 10.1002/cssc.201,
        pdfs.semanticscholar.org/b956/209528602aa65df010f76ffab0779c55f80c.pdf. Accessed 15 April   2018.
[14] Huang, Yao Bing. "Hydrolysis of cellulose to glucose by solid acid catalysts", *Royal Society of Chemistry*.   https://pubs.rsc.org/en/content/articlelanding/2013/gc/c3gc40136g#!divAbstract Accessed 28 Nov. 2018
[15] Wang, Ying. "Cellulose fibre dissolution in sodium hydroxide solution at low temperature: dissolution kinetics   and solubility improvements", *Georgia Institute of Technology*, December 2008.
        citeseerx.ist.psu.edu/viewdoc/download?doi=10.1.1.465.6848&rep=rep1&type=pdf. Accessed 21 April   2018.

Wool will be affected greatly in caustic soda solutions that would have little effects on cotton.[16]

Silk could be degraded more readily by acids than wool since it not only falls under the same category, but it also is a lot tender and delicate. Concentrated hot acids may dissolve the silk fibroin but dilute mineral or organic acids do not affect as significantly as it is affected by alkalis. [17]

5. Variables

Table 1: Variables

Type of variable	Name of variable	Description
Independent variable	Concentration of the nitric acid and sodium hydroxide solution used.	The concentrations of the mentioned solutions were incremented from 1 mol dm^{-3} to 5 mol dm^{-3}, by 1 mol dm^{-3}. NaOH solution was prepared using an electronic balance of uncertainty $\pm0.001g$, distilled water and a standard flask of 250 ml.
Dependent variable	The tensile strength of cotton, wool and silk.	The tensile strengths of the fibres soaked in the solutions were measured using a spring balance and standard weights. The error in the spring balance was measured to be $\pm0.33\%$.

[16] Mamun, Mustaque Ahammed. "Physical and chemical properties of wool fibre", *Textile learner*. textilelearner.blogspot.com/2015/12/physical-and-chemical-properties-of.html. Accessed 18 May 2018.

17 Khan, Rakibul Islam. "Physical and chemical properties of silk fiber". Textile Learner. Accessed 15 May 2018.textilelearner.blogspot.com/2013/06/physical-and-chemical-properties-of.html. Accessed 15 May 2018.

Controlled variables	Length of the fibre taken and the time for which it was soaked in the solutions.	The length of the thread taken was controlled at 14cm by using a simple ruler of uncertainty ±1mm and the soaking time was controlled at 5 minutes by the use of a digital stopwatch.

All procedures were carried out under standard temperature and pressure – 298 K and 101.6 kPa

6. Apparatus and materials

Table 2: List of apparatus

Name	Least count	Uncertainty
Electronic balance	0.001 g	Negligible
Standard flask – 250 ml	NA	NA
Spring balance	0.098 N	± 0.049N
Weights – Various	NA	NA
Ruler	1 mm	± 0.5 mm
Measuring Jar – 10 ml	1 ml	±0.5 ml
Measuring Jar – 50 ml	0.1 ml	±0.05 ml
Petri dishes	NA	NA
Stopwatch	0.01 sec	± 0.01 sec
Retort Stand	NA	NA

The chemicals used in the experiment were Nitric acid and Sodium hydroxide prepared at 5 mol dm^{-3}, which were later diluted according to requirements using distilled water.

7. Safety

Rubber gloves and goggles were worn while working with corrosive chemicals. Both nitric acid and sodium hydroxide are highly corrosive in nature, so care was taken so as to not let these affect any part of the apparatus, nor any person around it. Solutions of high concentrations were disposed off safely after dilution. Used fibres were washed and disposed into the trash and care was taken so as to not let them clog drains or sinks.

8. **Hypothesis**

Referring to section 4, it can be hypothesized that –

- The tensile strength of cotton will decrease on exposure to increasing concentrations of nitric acid but will remain almost constant when reacted with sodium hydroxide.
- The tensile strength of wool will decrease in both cases, but the curve will be deeper in the case of the alkali – because wool reacts to a greater extent with an alkali, than with an acid.
- The tensile strength of silk will decrease deeply in both cases, but the curve will be deeper in the case of the alkali. The curve of silk against the acid will be deeper than the one of wool.

9. **Method**

Part A – Controlling Variables

- Equal lengths measuring to 14 cm of the three fibres were cut out of which 2 cm on either side were set apart for tying. The 10 cm in the centre was soaked in a certain solution of a quantity of 20 ml. The length was measured using a 30 cm ruler.
- The timer was set at an upper limit of 5 minutes, so that every piece of fibre is soaked in either of the solutions for an equal time, to avoid discrepancy.

Part B – Procedure

- Nitric acid and Sodium Hydroxide solutions of concentration 5 mol dm^{-3} were prepared. These solutions were then diluted to concentrations of 1, 2, 3 and 4 mol dm^{-3}.

  $C_1V_1 = C_2V_2$ $\longrightarrow$ $5 \times 20 = 4 \times V_2$ $\longrightarrow$ $V_2 = 25$ cm^3

  This method was used to prepare solutions of various dilutions.
- The pieces of thread were then soaked in the solutions separately for 5 minutes each. These soaked fibres were then washed with distilled water and dried to avoid further unequal reaction between the fibre and solution.
- The tensile strength of each sample was measured. This was done by tying one end of the thread to the spring balance, the other end to the weight stand and gradually adding weights until the thread breaks. The mass of the weights was then recorded in grams and then converted to newton by multiplying the data with 0.0098 (One gram force = 0.0098 N).

- This was performed for all the fibres for all concentrations of both the solutions, and data was recorded thrice, for three different trials.

10. Results

10.1. Qualitative Data

When the fibres were soaked in the nitric acid, they turned pale yellow in colour, cotton and silk reacting the most. During reaction with sodium hydroxide solution, the fibres tend to separate as concentration increases, so the procedures had to be carried out very carefully, often including many numbers of trials. However, this effect decreased as the concentration was effectively decreased. The results of the experiments are shown below.

10.2. Quantitative data

Table 3: Cotton – Data values

Name	Tensile Strength of Cotton (in gram force) ($\pm$ 5gf)					
Solution	Sodium Hydroxide			Nitric Acid		
Concentration	Trial 1	Trial 2	Trial 3	Trial 1	Trial 2	Trial 3
1 mol dm^{-3}	1190	1180	1190	1110	1120	1140
2 mol dm^{-3}	1180	1190	1170	1030	1040	1040
3 mol dm^{-3}	1180	1170	1180	910	930	930
4 mol dm^{-3}	1180	1180	1160	750	720	740
5 mol dm^{-3}	1170	1160	1170	450	460	440

Table 4: Silk – Data values

Name	Tensile strength of Silk (in gram force) ($\pm$ 5gf)					
Solution	Sodium Hydroxide			Nitric Acid		
Concentration	Trial 1	Trial 2	Trial 3	Trial 1	Trial 2	Trial 3
1 mol dm^{-3}	450	440	450	490	480	490
2 mol dm^{-3}	390	380	390	470	460	470
3 mol dm^{-3}	260	250	260	410	430	430
4 mol dm^{-3}	70	80	90	360	370	380
5 mol dm^{-3}	NA	NA	NA	310	330	320

Table 5: Wool – Data values

Name	Tensile strength of Wool (in gram force) ($\pm$ 5gf)	
Solution	Sodium Hydroxide	Nitric Acid

Concentration	Trial 1	Trial 2	Trial 3	Trial 1	Trial 2	Trial 3
1 mol dm^{-3}	690	680	680	670	660	680
2 mol dm^{-3}	660	650	650	650	650	640
3 mol dm^{-3}	590	600	600	610	600	600
4 mol dm^{-3}	510	500	490	520	540	530
5 mol dm^{-3}	380	360	360	490	450	480

All the data values were recorded in gram force. These values were converted into newtons – by multiplying with 0.0098 - and averaged before being graphed.

Sample calculation-

Values of the reaction between NaOH (1 mol dm^{-3}) and cotton

$$\frac{1190 + 1180 + 1190}{3} \times 0.0098 = 11.63\ N$$

Table 6: Averaged and processed values in newton ($\pm$ 0.049 N)

Fibre	Cotton		Wool		Silk	
Solution	NaOH	HNO$_3$	NaOH	HNO$_3$	NaOH	HNO$_3$
1 mol dm^{-3}	11.63	11.01	6.70	6.57	4.38	4.77
2 mol dm^{-3}	11.60	10.19	6.41	6.34	3.79	4.57
3 mol dm^{-3}	11.53	9.05	5.85	5.91	2.52	4.15
4 mol dm^{-3}	11.49	7.22	4.90	5.19	0.78	3.63
5 mol dm^{-3}	11.43	4.41	3.59	4.48	NA	3.17

X-axis –	
Graph 1 – Concentration of NaOH in mol dm^{-3}.	Graph 2 – Concentration of HNO$_3$ in mol dm^{-3}.
Y-axis –	
Graph 1 and 2 – Tensile strength of cotton, silk and wool in newton.	

Graph 1 – Effect of NaOH on tensile strength of fibres

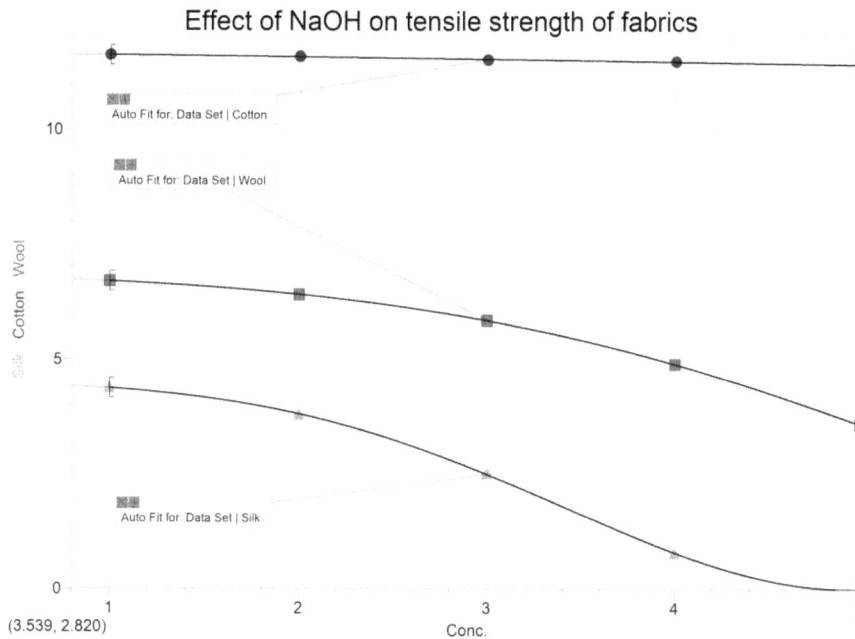

Effect of NaOH on tensile strength of fabrics

Source – Self, Self. Logger *Pro* 3.9

Graph 2 – Effect of HNO₃ on tensile strength of fibres

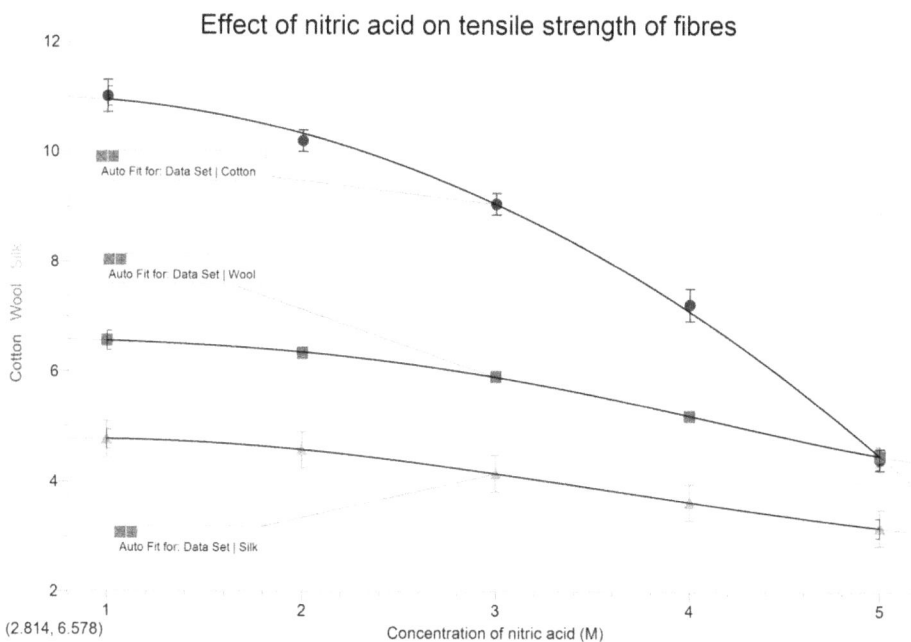

Effect of nitric acid on tensile strength of fibres

Source – Self, Self. Logger *Pro* 3.9

11. <u>Analysis and discussion</u>

The results prove that the hypotheses stated earlier are almost true. The tensile strength of cotton decreased more drastically in Graph 2 (acidic medium) than in Graph 1 (alkaline). In the alkaline medium cotton presented an almost linear relation with the strength decreasing with a slope of just -0.05, but in the case of the acid, the strength declined by a noticeable exponent of $e^{-0.4x}$. The decrease in the graph of cotton in the acidic medium is much greater than the other fibres, which do not fall as drastically when concentration is increased from 1 mol dm⁻³ to 5 mol dm⁻³. This is because the lignocellulose of cotton is more readily hydrolysed by an acid, than an alkali, in which the polysaccharides remain stable. The reverse can be observed with wool and silk, which remain fairly strong in acidic media (Graph 1) as compared to Graph 2, where the strength falls noticeably, with exponents of almost 2.

The tensile strength of wool also was affected as per the hypothesis. The tensile strength decreased almost quadratically in both the cases. The wool polymer is a linear alpha keratin polymer and made of polar peptide groups, linked via salt linkages, covalent peptide bonds and Van der Waal's forces. The peptide bonds are affected to a great extent when exposed to an alkaline medium, than an acidic one. The silk filament also behaved as expected, but for the curve in the acidic medium, whose slope turned out to be much steeper than anticipated. The silk polymer system is held together by salt linkages, hydrogen bonds and Van der Waal's forces of attraction. In an alkaline medium, the bonds of the silk polymer are all hydrolysed by the alkali, which leads to the weakening of the fibroin in the solution. Silk, being an amphoteric substance, might have also reacted with the nitric acid to a certain extent, which must have contributed to the depth in the curve.

The absolute error in the experiment cannot be expressed in terms of a relative percentage due to the lack of a literature value. This is because even if this investigation was done earlier, the results will be different due to the differences in the controlled variables – the length of the fibre taken, the area of cross-section of the thread and the time for which it was soaked. Nonetheless, the random error of the experiment contributed by the numerical uncertainty in measurement can be calculated.

Error Calculation

The uncertainty of the spring balance $= \pm 5g$

Sample calculation –

Percentage uncertainty in the value of the tensile strength of cotton in 1 mol dm^{-3} NaOH =

$$\frac{\frac{Range}{2}}{Mean\ Value} \times 100 \implies \frac{\frac{1190-1180}{2}}{1185} \times 100 = \pm 0.42\%$$

Table 6: Error in the experiment

Title	Error in the average value of tensile strength (in %)					
Medium	Sodium Hydroxide			Nitric Acid		
Fibre/ Conc.	Cotton	Wool	Silk	Cotton	Wool	Silk
1M	±0.42%	±0.42%	±0.42%	±1.26%	±0.84%	±0.42%
2M	±0.84%	±0.42%	±0.42%	±0.42%	±0.42%	±0.42%
3M	±0.42%	±0.42%	±0.42%	±0.84%	±0.42%	±0.84%
4M	±0.84%	±0.84%	±0.84%	±1.26%	±0.84%	±0.84%

5M	±0.42%	±0.84%	NA	±0.84%	±1.69%	±0.84%
Average	±0.59%	±0.59%	±0.52%	±0.92%	±0.84%	±0.67%

Total average percentage uncertainty

$$= \pm \frac{0.59+0.59+0.52+0.92+0.84+0.67}{6} \text{ N}$$

$$= \pm \mathbf{0.83\%}$$

Systematic error in the spring balance = 0.33%

Net uncertainty = ± (0.83 – 0.33) % = **±0.50%**

The ±0.33% error in the ruler was neglected because the effect in length was not being investigated, so it could not fit into any of the numerical error criteria.

12. Error evaluation

Table 7: Error evaluation

Error	Type of error	Effect on the final result	Solution
Error in the electronic balance	Systematic error	The increase or decrease in measuring the NaOH salt affects the concentration of the solution prepared, which in turn alters the reaction between the solution and fibre.	The salt must be weighed correctly twice or thrice, so as to decrease the risk of an error.
Error in the spring balance	Systematic error	The systematic error in the spring balance affects the reading of the tensile strength of the thread, thereby proving to be a major area to focus since it threatens the very foundation of this experiment.	This error can be minimized by checking the zero error and precision of the instrument prior to the experiment.
Error in time for which the thread was soaked in the solutions.	Random error	This error may or may not play an important role in the experiment, but still, it does play a role. The difference in duration can cause an unequal reaction in various samples thus again affecting the independent variable.	This error can be minimized by reducing the reflex time and paying attention to the stopwatch.
Error in washing or drying the thread	Random error	This error might have caused the discrepancies in the graphs – the tensile strength of a fibre depends greatly on whether the thread is dry or wet. For example, the tensile strength of cotton when wet is almost 1.5 times of what it is when it is dry.	This error can be minimized by checking properly whether the thread has dried or not, by dabbing a dry filter paper against it.

13. Conclusion

This experiment was conducted to investigate the effect of nitric acid and sodium hydroxide on cotton (plant fibre), silk and wool (animal fibres). From the above sections and the hypotheses, it can be concluded that the tensile strength of cotton, a plant fibre is greatly affected by HCl than by NaOH; and that the tensile strength of animal fibres is greatly affected by NaOH than by HCl. This can be verified from the following statements. The strengths of cotton, wool and silk in 1 mol dm^{-3} HNO_3 are 11.01, 6.57 and 4.77, respectively, and in 1 mol dm^{-3} NaOH are 11.63, 6.70 and 4.38 respectively. The strengths of cotton, wool and silk in 4 mol dm^{-3} HNO_3 are 7.22, 5.19 and 3.63 respectively, and in 4 mol dm^{-3} NaOH are 11.49, 4.90 and 0.78 respectively. The fall in tensile strength for the fibres in different pH conditions can be evidently observed from these statements. Overall, the strength of cotton remains higher in both the conditions, which in conclusion, gives a reason for such vast use of cotton in manufacturing laboratory wearables. The absolute error in the experiment could not be expressed in terms of a relative percentage due to the lack of a literature value.

The investigation can be further carried out by finding a function for the relationship between concentration and tensile strength. This will aid in estimating the tensile strength for any given concentration of solution, and not only those which were investigated. The relation between length of the fibre and tensile strength can also be investigated. The results of these researches may be of great use to the textile industry in the near future.

14. Bibliography

1. "Cotton Fiber Tech Guide", *Cotton Incorporated*, www.cottoninc.com/quality-products/nonwovens/cotton-fiber-tech- guide/ Accessed 31 March 2018.

2. Huang, Yao Bing. "Hydrolysis of cellulose to glucose by solid acid catalysts", *Royal Society of Chemistry*. pubs.rsc.org/en/content/articlelanding/2013/gc/c3gc40136g#!divAbstract Accessed 28 Nov. 2018

3. Mamun, Mustaque Ahammed. "Physical and chemical properties of wool fibre", Textile learner. textilelearner.blogspot.com/2015/12/physical-and-chemical-properties-of.html. Accessed 18 May 2018.

4. Rosa, Silvia Morales-dela, "Complete chemical hydrolysis of cellulose into fermentable sugars via ionic liquids and antisolvent pretreatments", ChemSusChem, DOI: 10.1002/cssc.201, Accessed 15 April 2018. pdfs.semanticscholar.org/b956/209528602aa65df010f76ffab0779c55f80c.pdf

5. Wang, Ying. "Cellulose fibre dissolution in sodium hydroxide solution at low temperature: dissolution kinetics and solubility improvements", Georgia Institute of Technology, December 2008.

citeseerx.ist.psu.edu/viewdoc/download?doi=10.1.1.465.6848&rep=rep1&type=pdf . Accessed 21 April 2018.

6. "Wool Fiber ‖ Physical and Chemical Properties of Wool", *Textile Fashion Industry*, 28 June 2012.

textilefashionstudy.com/wool-fiber-physical-and-chemical-properties-of-wool/ Accessed 23 Feb. 2018.

15. Software used

1. Logger*Pro* 3.9
2. MS Word Home and Student 2016.

6. HOW DOES THE TEMPERATURE OF AN ELECTROLYTE AFFECT ITS E VALUE AND THE EQUILIBRIUM POSITION?

Author: Sowmya Iyer
Moderated Mark: 23/24

1. BACKGROUND RESEARCH

Voltaic cells can be used to create a current through the flow of electrons. This occurs during a redox reaction between two metal electrodes and the electrolytes made of salts with their ions. This results in their being a shift in the shift in the equilibrium concentration, forcing the K_c to change because of the continuous reaction.[18]

Generating electricity is very important for our modern world, in which almost everything we do runs on electrical energy. With the growing depletion of nonrenewable sources of energy, we must begin to switch to alternative, more sustainable sources of renewable energy. By using redox reactions to create a current, we can generate electricity. Furthermore, by shifting the equilibrium position by changing the temperature of the solution, the amount of energy generated, measured by looking at the potential difference across the cell, can be increased for the same concentration of the electrolyte. For a voltaic cell, the electrolyte is a solution that contains the metal cations of the respective pure metal reactive electrode. A change in temperature of the electrolyte will shift the Kc value and hence, the optimal temperature can be determined.

While studying this chapter during chemistry, this was the first thing that came to my mind. If we can use reversible redox reactions that can produce direct currents, then electrolysis in a voltaic cell can be used to generate energy. However, while doing research on the topic, I realized that most of the research done was on the effect of the concentration of the electrolyte and the E^θ_{cell} value. This may pose an issue if during large-scale electrolysis; there is a lack of chemicals. However, the E^θ_{cell} value of an electrolyte is dependent on the temperature as well. This made me choose to research on this part of electrolysis, since not much temperature related research has been done. Furthermore, there was also very little research done on the Kc of the reaction. Since it is spontaneous, it constantly changes and hence, it is very hard to determine whether or not the temperature will affect the reaction based on trends established for normal equilibrium reactions.

2. THE RESEARCH QUESTION

The lack of temperature specific research on electrolysis made me come up with my research question, **how does the temperature of an electrolyte affect its E^θ_{cell} value and the equilibrium position?** The purpose of this experiment is to analyze how the temperature of the solution will affect the E^θ_{cell} value and hence, at what temperature can a higher P.D. be achieved.

The temperature will be varied in 10°C intervals ranging from 30°C to 90°C with a measurement at 25°C taken to compare to the value at Standard conditions. The temperature of only one electrolyte will be changed, so as to allow the analysis to be focused on just one electrolyte, the Cu^{2+} electrolyte, so that

---

[18] Bylikin, Sergey, Gary Horner, Brian Murphy, and David Tarcy. *Chemistry: Course Companion.* 2014 ed. N.p.: n.p., n.d. Print.

the Nernst equation can be used to calculate the Kc. The E^{θ}_{cell} value will be measured using a voltmeter probe that is connected to a data logger, so that any minor fluctuations can be taken into account and averaged out over a fixed period of time.

A Daniell Voltaic cell was used. The electrolysis equations are as follows.

Reduction half (at Cathode): Cu^{2+} (aq) + 2e$^-$ $\rightarrow$ Cu (s)

Oxidation half (at Anode): Zn (s) $\rightarrow$ Zn^{2+} (aq) + 2e$^-$

Full ionic equation: Cu^{2+} (aq) + Zn (s) $\rightarrow$ Cu (s) + Zn^{2+} (aq)

Independent variable: Temperature of Cu^{2+} electrolyte.

The temperature will be varied in 10°C intervals ranging from 30°C to 90°C with a measurement at 25°C taken to compare to the value at Standard conditions.

Dependent variable: E^{θ}_{cell} Value of Cu^{2+} electrolyte.

The E^{θ}_{cell} value will be measured to a precision of ±0.001V, using a voltmeter probe that is connected to a data logger, so that any minor fluctuations can be taken into account and averaged out over a fixed period of time.

Controlled variables:

1. Temperature of Zn^{2+} electrolyte, controlled by immersing the beaker in separate water baths with specific controlled temperatures.
2. Volume of electrolyte used, controlled by measuring the same volume of 70cm^3 for each electrolyte for every new set of readings.
3. Type of Electrode used, controlled by using electrodes of the same dimensions from the same original strip of metal.
4. Resistance caused by wires, controlled by using the same wires throughout the experiment, and waiting for intervals of 10 minutes between trials, so as to allow the wires to cool to room temperature to avoid an increase in resistance.
5. Time of electrolysis, controlled by starting the data logger before placing it into the solutions so that the reaction can be measured fro exactly (20.0 ± 0.2) seconds from the start of the reaction.

3. HYPOTHESIS

Using the Nernst equation, a predicted trend can be established. The equation is:

$$E = E^\theta - \frac{RT}{nF} \ln Q$$

Where Q is the equilibrium constant, calculated by $[Zn^{2+}]/[Cu^{2+}]$

Due to the Cu^{2+} being reduced and Zn being oxidized, the value of $[Zn^{2+}]/[Cu^{2+}]$ will increase, thus making $\ln Q$ increase. Also, the increase in the temperature will increase the overall value of $\frac{RT}{nF} \ln Q$ and hence, we expect the E value to decrease with an increase in temperature, thus producing a graph with a negative gradient.

4. CHEMICALS AND APPARATUS

Apparatus	Quantity	Uncertainty of instrument
100 cm^3 measuring cylinder	1	±0.5 cm^3
100 cm^3 beaker	4	±10 cm^3
Electronic mass balance	1	±0.001g
Water bath	2	±0.1°
Data logger	1	
±6V Voltmeter probe	1	±0.001V
Thermometer	1	±0.05°
5 x 1cm Copper strip	21	±0.3cm^2
5 x 1cm Zinc Strip	21	±0.3cm^2
Filter paper	21	
Petri dish	1	

Chemicals

1.0 mol dm^{-3} $CuSO_4$ solution

1.0 mol dm^{-3} $ZnSO_4$ solution

0.1 mol dm^{-3} KNO_3 solution
Distilled water

5. METHODOLOGY

1. Pour 70 cm^3 of each solution into separate 100 cm^3 beakers.
2. Set the temperature of one water bath at 25.0° C.
3. Set the temperature of the other water bath at 90.0° C.
4. Place the beaker containing $ZnSO_4$ solution in the water bath that is at 25.0° C.
5. Place the beaker containing $CuSO_4$ solution in the water bath that is at 90.0° C.
6. Pour some KNO_3 into the petri dish till it's about half full.
7. Create a salt bridge by rolling up a piece of filer paper and placing it into the petri dish to soak.
8. Cut the copper and zinc sheets into 21 strips that are of the dimensions 1 x 5 cm to make the electrodes.
9. Connect the voltage sensor to the data logger and set it to ±6 V.
10. Connect the positive terminal of the voltage sensor to the copper strip.
11. Connect the negative terminal of the voltage sensor to the zinc strip.
12. Place the copper strip into the $CuSO_4$ solution and the zinc strip into the $ZnSO_4$ solution.
13. Place each end of the salt bridge into each solution such that the tip is submerged but does not touch the electrode.
14. Press start on the data logger and record the potential difference across the terminals for 30 seconds.
15. Find and record the average value of the P.D. taking into account the fluctuations.
16. Repeat steps 1 - 16 two more times and take an average reading for the P.D.
17. Keep the temperature of the water bath for $ZnSO_4$ solution at a constant 25.0° C.
18. Repeat steps 1 - 18, while decreasing the temperature of the $CuSO_4 \bullet 5H_2O$ solution in increments of 10.0° C for every new set from 80.0° to 30.0° C.
19. Calculate the E_{cell}^{θ} Value of Cu^{2+} electrolyte by subtracting 0.76 from the measured P.D.
20. Plot a graph of E_{cell}^{θ} Value of Cu^{2+} electrolyte against the temperature of the solution.

6.   SAFETY ISSUES AND ENVIRONMENTAL IMPACTS

There are safety concerns and environmental impacts of voltaic cells. Firstly, the stability of the reaction mixture for any cell must be taken into account. For this experiment, the copper and zinc half-cells are rather stable and hence, it is safe to use. That being said, prolonged exposure to the metal may cause health issues and hence, the experiment should not be carried out for long periods of time. Furthermore, there is a possibility of galvanic corrosion of the half-cells. The corrosion of the metals can be dangerous and even lead to health issues, and hence, precaution must be taken to avoid this. Both the electrolytes and the electrodes can be changed between each set of experiments to avoid corrosion due to prolonged reaction.

If nitric acid is used during the production of copper sulfate, nitrogen dioxide gas will be produced, which is toxic. Hence, it must be done under the fume cupboard.[19] However, this is an expensive method and hence, electrolysis can be used to produce it, though hydrogen gas is the by-product and hence, the environmental impact of H_2 must also be taken into account. The production of zinc sulfate from the ore is also an expensive process that requires intensive heating at extremely high temperatures.[20] These high economic costs must be taken into account. However, producing electricity from a voltaic cell is one of the most environmental friendly ways, since it does not pollute the environment, and all the chemicals used can be re-used later in other experiments.

Furthermore, to increase the accuracy of the P.D. measured, the resistance of the wire will be reduced. This will allow a very high current to pass through the wires and this poses an issue of danger due to electrocution. The high current also makes the wires a fire hazard. Therefore, it is important to make sure that all the wires are thoroughly insulated and there are no live wires.

---

[19] "Make Copper Sulfate from Copper and Sulfuric Acid (3 Ways)."*Instructables.com*. N.p., n.d. Web. 03 June 2016. <http://www.instructables.com/id/Make-Copper-Sulfate-from-Copper-and-Sulfuric-acid-/>.

[20] "CIEC Promoting Science at the University of York, York, UK." *Zinc*. N.p., n.d. Web. 03 June 2016. <http://www.essentialchemicalindustry.org/metals/zinc.html>.

7. DATA COLLECTION AND PROCESSING

The temperature of each solution was measured in Celsius and then converted to Kelvin. The E_{cell}^{θ} of Cu^{2+} was calculated by subtracting the E_{cell}^{θ} of Zn^{2+} at SATP (0.76V).

Table 1. Raw and Processed Data to calculate P.D

Temperature/°C	Temperature/K	PD_1/V	PD_2/V	PD_3/V	PD_{avg}/V	E_{cell}^{θ} Cu^{2+}
±0.1°C	±0.1K		±0.001V		±0.001V	±0.001V
30.0	303.0	1.101	1.105	1.109	1.105	0.345
40.0	313.0	1.094	1.091	1.099	1.095	0.335
50.0	323.0	1.083	1.097	1.078	1.086	0.326
60.0	333.0	1.079	1.078	1.082	1.080	0.320
70.0	343.0	1.074	1.075	1.079	1.076	0.316
80.0	353.0	1.067	1.069	1.063	1.066	0.306
90.0	363.0	1.045	1.047	1.042	1.045	0.285

From these values, a graph of the E_{cell}^{θ} of Cu^{2+} against temperature can be plotted.

Graph of E_{cell} against Temperature

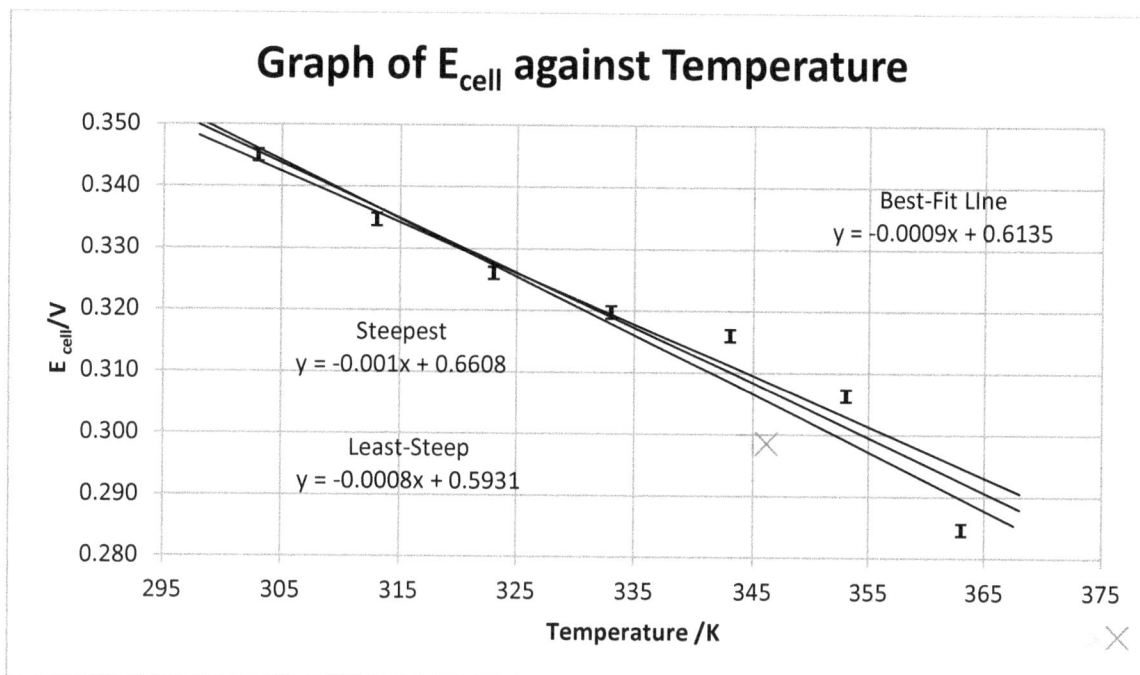

There is a linear relationship between the E^θ_{cell} of Cu^{2+} and the temperature of the electrolyte. The gradient if the graph is negative, showing that an increase in the temperature will result in a decrease in the E^θ_{cell} of Cu^{2+}. However, the magnitude of the gradient is of the order 10^{-4} and hence, it can be concluded that the change in the E^θ_{cell} of Cu^{2+} with respect to a small change in temperature is not extremely significant.

The y-intercept of the graph would show the predicted E^θ_{cell} of Cu^{2+} at 273.0K. The linear correlation coefficient is -0.974. The percentage deviation is 3%, which is within a 10% deviation from value of -1. This shows that the points on the graph are not too far off from the best-fit line and hence, the data can be considered precise.

However, what is most important about this equation is that we can substitute the temperature to be 298K and find the SATP value predicted by the trend. By substituting 298 as the temperature, we get the E^θ_{cell} of Cu^{2+} to be 0.345V (3 d.p.). The standard literature value for the E^θ_{cell} of Cu^{2+} at SATP is 0.34V[21]. Thus, the percentage error of the SATP value of the E^θ_{cell} of Cu^{2+} is $\frac{(0.345-0.34)}{0.34} \times 100\% = 1\%$ error.

Since the percentage error of the value is less than 10%, it can be considered accurate[22]. Hence, the values from the graph can be used in the Nernst equation to calculate the equilibrium constant at each value.

[21] IB Chemistry Data Booklet 2014
[22] "APPENDIX." *APPENDIX*. N.p., n.d. Web. 03 June 2016. <http://www.physics.mcmaster.ca/undergrad/Uncertainties/U.htm>.

Using the Nernst equation $E = E^\theta - \frac{RT}{nF} \ln Q$ the equilibrium constants at each temperature can be calculated. To calculate this, the equation can be rearranged such that: $\ln Q = -\frac{E - E^\theta}{\frac{RT}{nF}}$

The uncertainty for the E_{cell} value calculated from the graph can be shown below.

Calculation of Gradient and Vertical Intercept

For the best fit line:

Gradient = – 0.0009

Vertical intercept = 0.6135

For the extreme lines:

<u>Steepest Line:</u>   <u>Least-steep Line:</u>

Gradient = – 0.001   Gradient = – 0.0008

Δgradient = – 0.001 – (– 0.0009)   Δgradient = – 0.0009 – (– 0.0008)

        = – 0.0001   = – 0.0001

        = ±0.0001 (1 s.f.)   = ±0.0001 (1 s.f.)

Vertical intercept = 0.6608   Vertical intercept = 0.5931

Δvertical intercept = 0.6608 – 0.6135   Δvertical intercept = 0.6135 – 0.5931

        = 0.0473   = 0.0204

        = ± 0.05 (1 s.f.)   = ± 0.02 (1 s.f.)

Average uncertainty for gradient = (0.0001+0.0001)/2 = ±0.0001 (1 s.f.)

Percentage uncertainty of gradient = (0.0001/0.0009) x 100% = 11.1 %

= 10 % (1 s.f.)

Average uncertainty for vertical intercept = (0.05+0.02)/2 = ± 0.04 (1 s.f.)

Percentage uncertainty of vertical intercept = (0.04/0.6135) x 100% = 6.52%

= 7% (1 s.f.)

Equilibrium constants:

Detailed Calculation of Q	Detailed Calculation of ΔQ
1. Finding the PD at the specific temperature from the *graph of E$_{cell}$ of Cu^{2+} against Temperature*:	1. The uncertainty of the P.D. calculated through the *graph of E$_{cell}$ of Cu^{2+} against Temperature* can be found using the uncertainties of the gradient and the vertical intercept.
Equilibrium constant at 363K:	ΔE/E = Δgradient/gradient + Δintercept/intercept
E = -0.0009(363) + 0.6135	ΔE/E = (0.0001/0.0009) + (0.04/0.62)
= 0.2868	= 0.1763
= (0.29 ± 0.05) V	ΔE = 0.05 (1 s.f.)
2. Calculating lnQ using the Nernst equation:	2. Calculation the absolute uncertainty of lnQ using the relative uncertainty, calculated using the Nernst Equation:
lnQ = [0.2868 – 1.10]/[(-8.314 x 363)/(2 x 96500)]	ΔlnQ/lnQ = Δtemp/temp + ΔE/E
= 52.004	ΔlnQ/lnQ = 0.1/363.0 + 0.05/0.2868
= 52 ± 9	ΔlnQ/lnQ = 0.1746
	ΔlnQ = 9.08
	= ± 9 (1 s.f.)
	3. Propagation of uncertainty for simple equation *y = f(x)* :

3. Calculating the Equilibrium Constant, Q (Kc) from the lnQ value:	$\partial y = \frac{dy}{dx} \partial x = \frac{df(x)}{dx} \partial x$ $\Delta y = \frac{df(x)}{dx} \Delta x$ $y = ln(Q)$ $\Delta ln(Q)/ln(Q) = \frac{\Delta Q}{Q}$
$Q = e^{ln(Q)}$ $Q = 3.84695 \times 10^{22}$ $= 38 \times 10^{21} \pm 7 \times 10^{21}$	$\Delta Q = \Delta ln(Q)/ln(Q) \times Q$ $\Delta Q = (9/52)(3.84695 \times 10^{22})$ $= 6.658 \times 10^{21}$ $= 7 \times 10^{21}$
4. Calculation of Percentage Uncertainty: Uncertainty $= \frac{7 \times 10^{21}}{38 \times 10^{21}} \times 100\%$ $= 17.5\%$ $= 20\%$ (1 s.f.)	

Calculation of Q (Kc) and Propagation of uncertainties for each temperature:

Equilibrium constant at 353K:	$\Delta E/E$ = Δgradient/gradient + Δintercept/intercept
$E = -0.0009(353) + 0.6135$ $= 0.2958$ $= (0.30 \pm 0.05)$ V	$\Delta E/E = (0.0001/0.0009) + (0.04/0.62)$ $= 0.1763$ $\Delta E = 0.05$
Therefore, $lnQ = [0.2958 - 1.10]/[(-8.314 \times 353)/(2 \times 96500)]$ $= 53 \pm 9$	$\Delta lnQ/lnQ = \Delta temp/temp + \Delta E/E$ $\Delta lnQ/lnQ = 0.169$

Therefore, Q = 10 x 10^{22} ± 2 x 10^{22} Percentage uncertainty = (2 x 10^{22}/10 x 10^{22}) x 100% = 20 %	ΔlnQ = 8.95 = 9 ΔQ/Q = ΔlnQ/ln(Q) Therefore, ΔQ = 2 x 10^{22}
Equilibrium constant at 343K: E = -0.0009(343) + 0.6135 = 0.3066 = (0.31 ± 0.05) V Therefore, lnQ = [0.3066 – 1.10]/[(-8.314 x 343)/(2 x 96500)] = 54 ± 9 Therefore, Q = 28 x 10^{22} ± 5 x 10^{22} Percentage uncertainty = (5 x 10^{22}/28 x 10^{22}) x 100% = 20 %	ΔE/E = Δgradient/gradient + Δintercept/intercept ΔE/E = (0.0001/0.0009) + (0.04/0.62) = 0.1763 ΔE = 0.05 ΔlnQ/lnQ = Δtemp/temp + ΔE/E ΔlnQ/lnQ = 0.1616 ΔlnQ = 8.72 = 9 ΔQ/Q = ΔlnQ/ln(Q) Therefore, ΔQ = 5 x 10^{22}
Equilibrium constant at 333K: E = -0.0009(333) + 0.6135 = 0.3138 = (0.31 ± 0.06) V Therefore, lnQ = [0.3138 – 1.10]/[(-8.314 x 333)/(2x 96500)] = 55 ± 10 Therefore, Q = 8 x 10^{23} ± 1 x 10^{23} Percentage uncertainty = (1 x 10^{23}/8 x 10^{23}) x 100% = 20 %	ΔE/E = Δgradient/gradient + Δintercept/intercept ΔE/E = (0.0001/0.0009) + (0.04/0.62) = 0.1763 ΔE = 0.06 ΔlnQ/lnQ = Δtemp/temp + ΔE/E ΔlnQ/lnQ = 0.1915 ΔlnQ = 10.495 = 10 ΔQ/Q = ΔlnQ/ln(Q) Therefore, ΔQ = 1 x 10^{23}

Equilibrium constant at 323K:	$\Delta E/E$ = Δgradient/gradient + Δintercept/intercept
$E = -0.0009(323) + 0.6135$	
$= 0.3228$	$\Delta E/E = (0.0001/0.0009) + (0.04/0.62)$
$= (0.32 \pm 0.06)$ V	$= 0.1763$
	$\Delta E = 0.06$
Therefore, $\ln Q = [0.3228 - 1.10]/[(-8.314 \times 323)/(2 \times 96500)]$	$\Delta \ln Q/\ln Q = \Delta temp/temp + \Delta E/E$
$= 56 \pm 10$	$\Delta \ln Q/\ln Q = 0.1859$
Therefore, $Q = 21 \times 10^{23} \pm 4 \times 10^{23}$	$\Delta \ln Q = 10.382$
Percentage uncertainty $= (4 \times 10^{23}/21 \times 10^{23}) \times 100\%$	$= 10$
	$\Delta Q/Q = \Delta \ln Q/\ln(Q)$
$= 20\%$	Therefore, $\Delta Q = 4 \times 10^{23}$
Equilibrium constant at 313K:	$\Delta E/E$ = Δgradient/gradient + Δintercept/intercept
$E = -0.0009(313) + 0.6135$	
$= 0.3318$	$\Delta E/E = (0.0001/0.0009) + (0.04/0.62)$
$= (0.33 \pm 0.06)$ V	$= 0.1763$
	$= 0.06$
Therefore, $\ln Q = [0.3318 - 1.10]/[(-8.314 \times 313)/(2 \times 96500)]$	$\Delta \ln Q/\ln Q = \Delta temp/temp + \Delta E/E$
$= 57 \pm 10$	$\Delta \ln Q/\ln Q = 0.1859$
Therefore, $Q = 6 \times 10^{24} \pm 1 \times 10^{24}$	$\Delta \ln Q = 10.321$
Percentage uncertainty $= (1 \times 10^{24}/6 \times 10^{24}) \times 100\%$	$= 10$
	$\Delta Q/Q = \Delta \ln Q/\ln(Q)$
$= 20\%$	Therefore, $\Delta Q = 1 \times 10^{24}$
Equilibrium constant at 303K:	$\Delta E/E$ = Δgradient/gradient + Δintercept/intercept
$E = -0.0009(303) + 0.6135$	
$= 0.3408$	$\Delta E/E = (0.0001/0.0009) + (0.04/0.62)$
$= (0.34 \pm 0.06)$ V	$= 0.1763$
	ΔE = 0.06

Therefore, $\ln Q = [0.3408 - 1.10]/[(-8.314 \times 303)/(2 \times 96500)]$ $= 58 \pm 10$ Therefore, $Q = 15 \times 10^{24} \pm 3 \times 10^{24}$ Percentage uncertainty $= (3 \times 10^{24}/15 \times 10^{24}) \times 100\%$ $= 20\%$	$\Delta \ln Q/\ln Q = \Delta temp/temp + \Delta E/E$ $\Delta \ln Q/\ln Q = 0.1764$ $\Delta \ln Q = 10.230$ $= 10$ $\Delta Q/Q = \Delta \ln Q/\ln(Q)$ Therefore, $\Delta Q = 3 \times 10^{24}$
Equilibrium constant at 298K: $E = -0.0009(298) + 0.6135$ $= 0.3453$ $= (0.35 \pm 0.06)$ V Therefore, $\ln Q = [0.3453 - 1.10]/[(-8.314 \times 298)/(2 \times 96500)]$ $= 59 \pm 10$ Therefore, $Q = 42 \times 10^{24} \pm 7 \times 10^{24}$ Percentage uncertainty $= (7 \times 10^{24}/42 \times 10^{24}) \times 100\%$ $= 20\%$	$\Delta E/E = \Delta gradient/gradient + \Delta intercept/intercept$ $\Delta E/E = (0.0001/0.0009) + (0.04/0.62)$ $= 0.1763$ $\Delta E = 0.06$ $\Delta \ln Q/\ln Q = \Delta temp/temp + \Delta E/E$ $\Delta \ln Q/\ln Q = 0.1741$ $\Delta \ln Q = 10.271$ $= 10$ $\Delta Q/Q = \Delta \ln Q/\ln(Q)$ Therefore, $\Delta Q = 7 \times 10^{24}$

Table 2. Processed Data to calculate Equilibrium Constant, Kc

Temperature / K	ΔTemperature / K	Kc / 1×10^{21}	ΔKc / 1×10^{21}
363.0	0.1	38	7
353.0	0.1	100	20
343.0	0.1	280	50

333.0	0.1	800	100
323.0	0.1	2100	400
313.0	0.1	6000	1000
303.0	0.1	15000	3000
298.0	0.1	42000	7000

The general trend shows a decrease in the equilibrium constant with an increase in temperature. A graph of the equilibrium constant against the temperature was plotted.

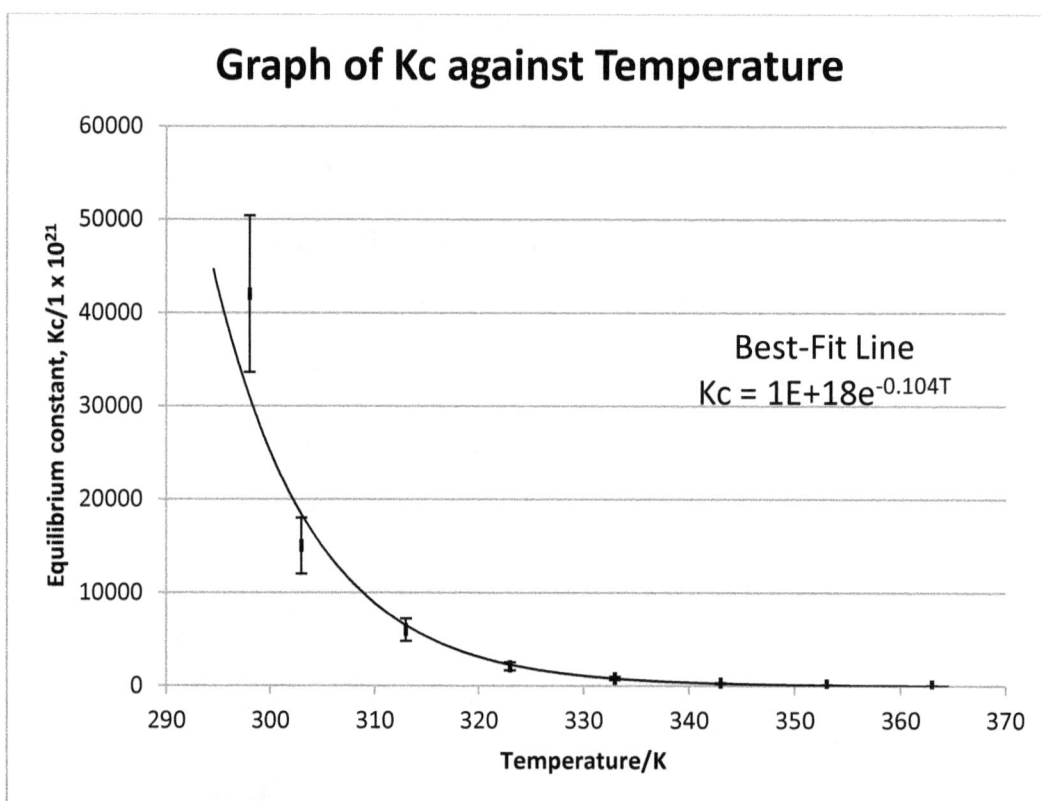

Graph of Kc against Temperature

Best-Fit Line

$Kc = 1E{+}18e^{-0.104T}$

It can be seen that there is an exponential relationship between the temperature and the equilibrium constant. Being an exothermic reaction, the Kc is expected to decrease with an increase in temperature. Therefore, it can be concluded that the shape of the graph follows theory. The correlation coefficient, R is -0.994, which shows that the best-fit line is precise and passes through almost all the points. However, as can be seen from the graph, the error bars for the Kc are significant. The percentage uncertainty for each individual point was calculated under the propagation of uncertainties. The percentage uncertainty for each of the points is approximately 20%. Thus, due to the fact that the percentage error is more than 10%[23], the Kc values obtained are limited in precision.

The Nernst equation can be used to calculate the theoretical Kc value at standard states, to compare to the experimental value to find the percentage error. At equilibrium, the E will be 0 as there is no net flow of electrons. However, since the reaction is spontaneous, this value will only be temporary, and hence is limited in its accuracy. Thus, the E value at SATP will be used to calculate the Kc value to compare for accuracy. A very high Kc indicates that the reaction is essentially complete.[24]

Therefore, $0 = E^{\theta}_{cell} - E - \dfrac{RT}{nF} \ln Q$

At standard states, $0 = (1.10 - 0.34) - [((8.314 \times 298)/(2 \times 96500))(\ln Q)]$

$\ln Q = (1.10 - 0.34)/[(8.314 \times 298)/(2 \times 96500)]$

Therefore, $\ln Q = 59.2$

$$Q = 5.1474 \times 10^{25}$$

$$Q = 5.15 \times 10^{25} \text{ (3 s.f.)}$$

The experimental value calculated from the graph at is 298K is 31×10^{24}

The percentage error is $= (51.5 \times 10^{24} - 31 \times 10^{24})/ (5.15 \times 10^{25}) \times 100\%$

$= 39.806$

$= 40 \%$ (1 s.f.)

8. CONCLUSION AND EVALUATION

From this experiment, the effect of temperature on both the E^{θ}_{cell} of the Cu^{2+} electrolyte and the equilibrium constant for the Cu^{2+} (aq) + Zn (s) → Cu (s) + Zn^{2+} (aq) reaction can be found. It can be seen that an increase in the temperature of the electrolyte will decrease its E^{θ}_{cell} value. Hence, to generate more electricity, a lower temperature is preferred.

The percentage uncertainty of the E^{θ}_{cell} value was calculated to be between 7-10% due to the individual uncertainties of the gradient and the vertical intercept. Thus, the data obtained can be considered precise. The percentage uncertainty for the Kc value was 20% for each of the data points. Thus, since the percentage uncertainty if higher than 10%, the data is limited in precision.

[23] "APPENDIX." *APPENDIX*. N.p., n.d. Web. 03 June 2016. <http://www.physics.mcmaster.ca/undergrad/Uncertainties/U.htm>.
[24] "K_Values_Nernst." *K_Values_Nernst*. N.p., n.d. Web. 03 June 2016. <http://www2.ucdsb.on.ca/tiss/stretton/CHEM2/K_Values_Nernst.html>.

The percentage error of the E^{θ}_{cell} value was calculated to be 1%. Therefore, the value can be considered accurate and hence, it can be assumed that the graph is accurate enough for further calculations. However, the percentage error of the Kc at SATP is 40% and hence, it is limited in accuracy.

Therefore, it can be concluded that, due to the limited accuracy in the value of the Kc, the data is not extremely reliable and a reliable conclusion can only be drawn based on the general trend.

Strengths and Weaknesses

The experimental procedure for the measurement of the potential difference across the two half-cells was effective. The results were both accurate and precise, thus making them reliable to use for further investigation. The experimental results could be used for further application, thus increasing the practical significance of the research in the field. The range of values measured was limited, hence limiting the reliability of the applications of the conclusions to a wider set of data.

There are some sources of error that can contribute to both accuracy and the precision of the data. Procedural errors will account greatly for the errors in the experimental values.

Sources of Error

1. The most important error in the Kc value was the fact that the Kc was not measured exactly at the point of dynamic equilibrium. This results in there being a high random error for the experiment. Therefore, the Kc values calculated were average values over a fixed amount of time. Though the amount was short, the reaction, being spontaneous, will see a rapid shift in the Kc value and hence, the calculated value will definitely be lower that the expected value.

2. The measured data is limited by the precision of its measuring instruments, resulting in random error. The precision of the data logger is 0.001V. However, the probe used measured ±6V. Though the number of decimal places measured remained the same, the precision of the data recorded was reduced due to the larger range.

3. The range of temperatures used for the experiment as well as the size of the interval also caused random errors. The change of temperature of the electrolytes was within 65°. The temperature was increased in 10° intervals. Therefore, due to the small range and interval size, the data was not very accurate.

4. Another major source of both random and systematic error came from the heating of the solutions in a water bath. Though a thermometer was used to measure the temperature of the solution before the start of the experiment, the temperature of the electrolyte may not remain constant throughout the course of the experiment. Though water has a high specific heat capacity and hence will not drop in temperature very fast, there will be an inevitable loss of heat to the surroundings during the experiment and hence, this will also result in errors of both the E^{θ}_{cell} value as well as the Kc.

5. The resistance in the wires will also cause a systematic error. The high resistance of the wire will result in the measured value of the P.D to be inaccurate, thus affecting both the E^{θ}_{cell} value as well as the Kc.

Reduction of Errors

1. Unfortunately, this error cannot be completely corrected as the instant the reaction starts, the equilibrium position will begin to shift. However, using the measured E value at the very first second can reduce it. The precision of the data logger must be adjusted for this. The number of samples taken per second can be increased so that there is still a large sample size to average out for an accurate reading. The smaller amount of time will allow the equilibrium position to be higher, and hence, more accurate.

2. A voltmeter probe of a smaller range can be used to measure the potential difference across the two half-cells. The 6V probe can be used to determine the predicted value for each of the temperatures and then a voltmeter with a more precise range can be used to increase the overall precision of the readings.

3. The range of temperatures at which the experiment was carried out should be increased. The size of the intervals between each of the readings should also be increased. This will allow the observed change in the P.D. to be more significant, thus making the trend clearer. Furthermore, the increase in the size of the intervals will help account for the value of the P.D. that lies slightly away from the trend line.

4. To reduce the error cause by the loss of heat to the surroundings, both water baths can be covered, and further insulated so that the heat is trapped within. This will reduce the heat that leaves the immediate surroundings of the electrolytes, thereby reducing the difference in temperatures, thus reducing the net energy transferred through heat. Additionally, the solutions can be left in the water bath for a prolonged period of time such that the entire solution is at the desired temperature.

5. Wires with a lower resistance can be used, such as copper wires. The diameter of the wire can also be increased to reduce the overall resistance. Also, the length of the wire can be reduced such that the circuit is complete with minimal wire. This will also minimize the total resistance in the circuit. This will increase the accuracy of the measured E_{cell}^{θ} value and hence, the calculated Kc value will be more accurate. However, safety precautions must be taken into account due to the high current present in the wires due to the lower resistance.

Extension

The purpose of this experiment was to determine the relationship between the temperature of an electrolyte and the E_{cell}^{θ} value as well as the Kc. Seeing as this experiment was limited in accuracy, the sources of error can be reduced so as to increase the accuracy of the results and hence, increase the reliability of the observed trend. The experiment can be repeated with different electrolytes so as to see if the trend hold true. Furthermore, the experiment can be carried out with endothermic reactions so as to see if the trend is the opposite due to the direction of shift of the Kc. The concentration of the electrolytes can also be increased and the temperature can be changed again to see if the extent of change of the Kc is greater. Based on that, the optimum relationship between the concentration and the temperature of the electrolyte can be determined, so that the highest P.D. is generated. This can be used to increase the efficiency of energy production, since electrolysis is an environmentally friendly method.

7. INVESTIGATING THE DISSOCIATION OF CALCIUM SUBSTANCES IN WATER IN THE PRESENCE OF CARBON DIOXIDE.

Author: Adam Zhou
Moderated Mark: 22//24

As the concentration of carbon dioxide in the atmosphere increases, it would drive the equilibrium to the right to counteract this change in accordance to Le Chatelier's Principle. Because of the large increase of carbon dioxide in the atmosphere, more of it is dissolved into the ocean, which drives Equations 2, 3, and 4 forward producing a larger proportion of H^+ and carbonate ions. That is, the excess proton acceptors in the seawater, or excess base, are able to slow down changes in pH given its affinity to interact with other various rapid coupled pairs of acids and bases occurring in the water and hence maintain overall system at equilibrium (Carpenter, 2014). Thus, there will be no drastic changes in the pH, and even when CO_2 is added, the number of positive ions (e.g. H^+), through the buffer solution, will be equivalent to the number of negative ions (e.g. carbonate ions) through the interactions with acids and bases (Carpenter, 2014). In standard, healthy seawater, there's enough CO_2 present in the atmosphere, and the buffer is maintained at a slightly basic pH of approximately 8.2 (Dockrill, 2019).

With my background and interest in environmental sustainability research, such as in being an ambassador for Samsung Engineering's environmental networking program, I have been exposed to how atmospheric carbon dioxide has risen from 280 parts per million (ppm) in preindustrial levels to current levels (2019) of 415 ppm (Dockrill, 2019). This in regards to anthropogenic ocean acidification that was mainly observed by climate scientists from reasons such as but not limited to carbon dioxide emissions from factories and vehicles. Though the buffer is still present, with the decreased total alkalinity being replaced with excess carbon dioxide, there has been a decrease in 0.11 pH since the industrial revolution (Dockrill, 2019). In this scenario, the hydrogen ion concentration also increases proportionally to the buffer solution's constituents of the hydrogen carbonate ion and the carbonate ion. Given that hydrogen ion concentration is associated by logarithmic factors of 10, this 0.11 pH is equivalent to a 29% increase.

There are an abundance of marine life that are affected, but one major impact are the fauna with calcium compounds in seawater. This is mainly calcium carbonate ($CaCO_3$) which are found in corals and shells used for protection. Increased pH from the former products of H^+ reacting with calcium compounds can lead to these shells softening because of the reaction:

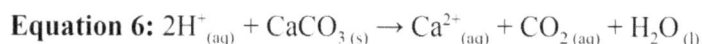

$$\textbf{Equation 6: } 2H^+_{(aq)} + CaCO_{3\,(s)} \rightarrow Ca^{2+}_{(aq)} + CO_{2\,(aq)} + H_2O_{(l)}$$

This reaction is a very spontaneous reaction favoring the forward direction, hence it is not at equilibrium. It is established that the reaction between carbonates and acids is very reactive with an equilibrium constant very much greater than one. The production of CO_2 will further increase the concentration of dissolved CO_2 in seawater. From Equation 2, more dissolved CO_2 will further drive the reaction forward, increasing the H^+ concentration, and resulting in more $CaCO_3$ breakdown.

There are other parameters that this investigation will not specifically address but are nevertheless important components of this system, i.e. salinity, temperature, and pressure, each with their own equilibrium constants. Thus, Le Chatelier's principle will only be able to identify the equilibrium's response rather than the net effect of the other aforementioned components. However, we are able to quantify this by measuring the direct product of calcium ions in the solution as the final dissolved product. This leads to the research question: **What is the relationship between the concentration of carbon dioxide (ppm) diffusion in a seawater solution and the concentration of Ca^{2+} ions (M) dissolved after a period of 24 hours at constant room temperature (25 °C) and pressure (1 atm)?**

2. Investigation

Hypothesis

Given a period of 24 hours, the calcium ions concentration will increase given a higher concentration of carbon dioxide diffused in the seawater. This assertion is the H_1 or the alternative hypothesis which states that there is a significant difference between these two variables. Then, the H_0 or null hypothesis value is that there is no significant difference between these two variables.

Variables

Independent Variable: Concentration of CO_2 diffused in the seawater, measured in ppm. This will be varied through a valve connected to a carbon dioxide tank with interval settings for the efflux of the gas. To find the concentration of ppm of carbon dioxide released into the solution, a titration procedure would be needed. However, it was found in the manual of the Harris HP721AL-500-705 Pressure Valve that each pressure level setting is equivalent to 10 ppm when diffused in water (e.g. level 3 will be equivalent to 30 ppm). Hence, we will use this assumption for this investigation and this limitation will be addressed in the evaluation. In regards to the error of the ppm calculation, another assumption has to be made. Given how the significant figures are set to 20 instead of 20.0, it can be assumed that there error of the pressure valve is ±1. Hence, we could use a quantitative independent variable.

Dependent Variable: Concentration of Ca^{2+} ions in the solution, measured in Molars (M) through a titration with the chelating agent EDTA. The production of these ions is due to the reaction of calcium carbonate with H^+ is not considered to be a very reversible reaction as the reaction proceeds in a forward direction spontaneously. Through such a titration, the EDTA is able to react with the individual ions in the solution by attaching its two or more donor atoms to each ion. Thus, it will form a more stable complex that will be able to be identified in the titration itself ("EDTA Titration"). This was chosen to be the dependent variable since seawater equilibrium tends towards the direct production of these ions during the dissolving of calcium carbonate. The production of the Ca-EDTA complex will not result in the production of more calcium ions as no additional H^+ ions are being introduced into the titration reaction and no solid $CaCO_3$ is present during the titration to provide additional calcium ions into the solution.

Control Variables:
- The same concentration of initial seawater and EDTA. By sourcing the seawater from the same place, and preparing the same procedure of making the EDTA solution, it will ensure that the calculations made to calculate for Molarity through $M_1V_1 = M_2V_2$ calculations. To solve for the unknown concentration within the same independent variables, that is, of calcium ions, then the other values should be kept constant.
- Same volume of initial seawater for each experiment, with the same principles as above.
- Same mass of other constituents in EDTA solution. Specifically, only 0.05 g of magnesium chloride hexahydrate will be used to allow for a sharper visual enhancement of the endpoint by forming a more stable complex with Eriochrome Black T indicator. Meanwhile, specifically only three pellets of sodium hydroxide will allow for the precipitation of magnesium chloride hexahydrate ("EDTA Titration").
- Same pH of the buffer solution for indicator measurement. This will be set at pH 10 as to allow the metallochromic indicator Eriochrome Black T to show a color change in the presence of a metal ion through the complexometric titration with EDTA. At pH 10, it allows for the EDTA to be present in a monoprotonated form and thus dissociate from the complex it makes with metal ions ("EDTA Titration").
- The time of day when readings were taken. This ensures that the time taken between the start of the experiment and the end of the diffusion of the carbon dioxide is 24 hours only, and no excess dissolving of $CaCO_3$ will take place.
- All corals were sourced from the same place, at Jepoi Capati's Lab so that the abiotic conditions they were in beforehand were as similar as possible.
- Same mass of coral pebbles with similar sizes and masses especially regarding similar surface area. A larger surface area of coral may induce more dissolving of calcium ions by being in contact with the carbon dioxide diffusion and thus induce an inaccurate result. Though these calculations will not be quantified and compared with one another, estimates will be made, and this limitation will be discussed in the evaluation.
- Same temperature and atmospheric pressure conditions. In accordance with the Ideal Gas Law, the diffusion of carbon dioxide into a solution should be kept in the same temperature and pressure to prevent the occurence of carbon dioxide moles evolved. However, this is kept fairly constant since environmental conditions remain as such. During the experiment, this is fairly straightforward, yet during storage, seawater and the chemicals should be closed and kept in a dark dry area.

Apparatus

Ethylenediaminetetraacetic acid (EDTA)

Eriochrome Black T Indicator

Sodium Hydroxide Solid Pellets

Anhydrous Magnesium Chloride Hexahydrate Solid

Sodastream 60L Carbon Dioxide Tank

Harris HP721AL-500-705 Pressure Valve for Carbon Dioxide Tank

NH_4Cl Buffer Solution (pH 10)

Seawater

25 pieces of approx 10g coral

25 small transparent tubs (7.5cm radius, 9cm height)

Measuring scale (±0.01 g)

5 250mL Erlenmeyer flasks (± 0.1 mL)

1 250 mL beaker (± 0.1 mL)

1 500 mL beaker (± 0.1 mL)

100 mL graduated cylinder (± 0.5 mL)

25 mL graduated cylinder (± 0.1 mL)

1 10 mL pipette and pipette pump (± 0.02 mL)

1 50 mL burette and burette clamp (± 0.05 mL)

Deionized water

Metal spatula

Stirring rod

Setup

Note: Bottle is replaced with a larger tub in this experiment. Corals are placed inside of the tub.

Procedure

The following procedure was designed by myself:

1. Set up the seawater sample by obtaining 500 mL through multiple measurements in the 100 mL graduated cylinder and pour it into a plastic tub.
2. Now with a mass balance, measure approximately 20 grams of a singular coral piece and place into the seawater sample.
3. After 24 hours, take a sample 20 mL aliquot of the seawater and store inside an Erlenneyer flask. This will act as the control measurement.
4. Attach pressure valve to carbon dioxide tank and place tubing into the seawater sample.
5. Set pressure valve at minimum setting.
6. Release the safety knob, noting the time at which the gas is first released.
7. After 24 hours, close the safety knob to stop the continuous gas flow.
8. Take a sample 20 mL aliquot of the seawater and store inside an Erlenneyer flask.
9. Repeat steps 1-9 for the other independent variables, but changing the pressure settings set at the intervals (level 2, level 2, level 3, and level 4) which are marked on the valve.
10. Prepare the EDTA sample by weighing 2g of reagent grade disodium EDTA in a 250 mL beaker. Add 0.05 g magnesium chloride hexahydrate, three pellets of NaOH. Add about 120 mL of distilled water using a 25 mL graduated cylinder. Stir until fully dissolved. This will produce a 0.05 M concentration of EDTA.
11. Using repeated measurements of a pipette and pipette pump, measure 50 mL of EDTA solution and place into a burette.
12. Take the seawater sample obtained from steps 3 and 9 and add 3 mL ammonium chloride buffer (pH 10) and 2-3 drops of Eriochrome Black T indicator solution.
13. Titrate with EDTA while stirring the seawater sample and observe a color change from violet through wine-red to blue to find the endpoint.

Risk Assessment

Ethical: The use of coral in this experiment may not have been ethically sourced depending on whether or not it came from live specimens, or dead ones that have been already washed up on shore. Note will be made to choose those coming from the latter to avoid loss in biodiversity and a potential effect on equilibrium of flora and fauna systems in the marine ecosystem.

Safety: The release of carbon dioxide is hazardous if in high amounts over 5000ppm, especially without proper ventilation and if concentrated in a small room for the experiment, leading to possibly nausea and headaches ("OSH Answers Fact Sheets"). Note will be made to conduct the experiment under a fume hood. Furthermore, the chemicals such as EDTA or NH_4Cl may cause skin irritation upon contact. If this is the case, rinse with water immediately. Wearing gloves and safety goggles are also imperative.

Environmental: The disposal of the aforementioned chemicals must be disposed of properly to avoid bioaccumulation of toxins in the food web and possible eutrophication. Diluting concentrated substances should be noted before directly pouring into the drain.

3. Data Presentation

Table 1: Raw Data Table Showing the Relationship Between the Amount of Diffused Carbon Dioxide in Seawater and the Initial and Final Volume of EDTA where Volume of Seawater of 20 mL and Concentration of EDTA of 0.05 M are Constant

Amount of Diffused Carbon Dioxide in Seawater (ppm ±1 ppm)	Initial Volume of EDTA (mL ± 0.05 mL)					Final Volume of EDTA (mL ± 0.05 mL)				
	Trial 1	Trial 2	Trial 3	Trial 4	Trial 5	Trial 1	Trial 2	Trial 3	Trial 4	Trial 5
10	3.40	10.90	17.90	25.10	32.60	10.90	17.90	25.10	32.60	40.50
20	4.10	13.70	23.60	32.80	5.60	13.70	23.60	32.80	42.20	15.30
30	15.30	7.80	26.90	10.20	29.80	34.60	26.90	45.90	29.80	49.30
40	5.20	2.60	3.30	2.10	6.00	47.80	45.20	46.20	44.30	48.10
50	2.50	1.70	1.60	2.20	0.90	80.40	87.10	86.90	88.40	87.30

Table 2: Processed Data Table Showing the Relationship Between the Amount of Diffused Carbon Dioxide in Seawater and the Change in Volume of EDTA Needed to Titrate an Unknown Concentration of Seawater, where Volume of Seawater of 20 mL and Concentration of EDTA of 0.05 M are Constant

Amount of Diffused Carbon Dioxide in Seawater (ppm ±1 ppm)	Change in Volume of EDTA (mL ± 0.1 mL)				
	Trial 1	Trial 2	Trial 3	Trial 4	Trial 5
10	7.5	7.0	7.2	7.5	6.9
20	9.6	9.9	9.2	9.4	9.7
30	19.3	19.1	19.0	19.6	19.5
40	42.6	42.6	42.9	42.2	42.1
50	85.9	85.4	85.3	86.2	86.4

Qualitative Observations:

During the reaction, it was noted that the precipitate continuously formed at the bottom of the solution. This precipitate was a brown-gray color and was noticeably more prominent with a higher ppm of carbon dioxide diffusion. Starting from 40-50 ppm as the independent variables, a slight effervescence and fizzing was observed in the water, distinct from the diffuser bubbling.

Processed Data

Given the formula of $Ca^{2+} + C_{10}H_{16}O_8N_2^{4-} \rightarrow C_{10}H_{12}O_8N_2Ca^{2-}$, it is identified how there is a 1 to 1 mole ratio between the calcium ions and the EDTA molecule. To calculate for the unknown concentration, apply the formula $M_1V_1 = M_2V_2 = n$

where M_1 and V_1 refer to the concentration (M) and volume (mL) of EDTA respectively, and M_2 and V_2 refer to the concentration (M) and volume (mL) of seawater (calcium ions) respectively. n is the number of moles and given the mole ratio of 1 to 1 established earlier, this equation is true.

Given M_2 is the unknown variable, we can transpose the equation to the following:

$$M_2 = \frac{M_1V_1}{V_2}$$

Substituting values, for example, in Trial 1 for 10 ppm Independent Variable:

$$M_2 = \frac{0.02 \times 0.05}{0.0075} = 0.0188$$

Table 3: Processed Data Table Showing the Relationship Between the Amount of Diffused Carbon Dioxide in Seawater and the Concentration of Seawater

Amount of Diffused Carbon Dioxide in Seawater (ppm ±1 ppm)	Concentration of Calcium Ions in Seawater (M)				
	Trial 1	Trial 2	Trial 3	Trial 4	Trial 5
10	0.0188	0.0175	0.0180	0.0188	0.0173
20	0.0240	0.0248	0.0230	0.0235	0.0243
30	0.0483	0.0478	0.0475	0.0490	0.0488
40	0.107	0.107	0.107	0.106	0.105
50	0.215	0.214	0.213	0.216	0.216

Error Calculations for Equipment Uncertainties from Table 3

1. To calculate for percentage uncertainty of concentration of EDTA, get the percentage uncertainties of both mass and volume given the formula for concentration is $\frac{\frac{mass}{molar\,mass}}{volume}$ (molar mass is not factored since it does not have an uncertainty).

Formula for both mass and volume percentage uncertainties are the following:
$\frac{equipment\,uncertainty}{calculated\,mass} \times 100\%$ and $\frac{equipment\,uncertainty}{calculated\,volume} \times 100\%$ respectively

Substituting values from Table 1:

Mass: $\frac{\pm 0.01 g}{2 g} \times 100\% = 0.50\%$

Volume: $\frac{\pm 0.1 mL}{120 mL} \times 100\% = 0.08\%$

Adding these values yields the percentage uncertainty of concentration: $0.50\% + 0.08\% = 0.58\%$

2. To calculate the percentage uncertainty of the volume of seawater, apply the formula above:

Substituting values: $\frac{\pm 0.1 mL}{20 mL} \times 100\% = 0.50\%$

3. To calculate the percentage uncertainty of the volume of EDTA, apply the same formula for each individual trial. For a sample, we will use trial 1 of the 10 ppm independent variable:

Substituting values: $\frac{\pm 0.1 mL}{7.5 mL} \times 100\% = 1.33\%$

4. To calculate percentage uncertainty of the concentration of calcium ions, apply the formula:

%uncertainty of concentration of EDTA + %uncertainty of volume of seawater + % uncertainty of volume of EDTA

Substituting values (trial 1 of 10 ppm independent variable): $0.58\% + 0.50\% + 1.33\% = 2.4\%$

5. To calculate the absolute uncertainty of the concentration, apply the following formula

Percentage uncertainty × Calculated value of Concentration of Seawater (Calcium Ions)

Substituting values for trial 1 for 10 ppm Independent Variable: $0.0241 \times 0.0188 = 0.00045$

For purposes of visual data presentation, the standard deviation was chosen as the error bars as in the context of this experiment, it had a higher error. Hence, sig figs are adjusted according to the decimal points present in standard deviation.

Table 4: Processed Data Table Showing Average Concentrations of Seawater Given Amount of Diffused Carbon Dioxide in Seawater Including Standard Deviation and Absolute Uncertainties

Amount of Diffused Carbon Dioxide in Seawater (ppm ±1 ppm)	Average Concentration of Calcium Ions (M)	Standard Deviation	Absolute Uncertainties
10	0.0188	0.0007	0.000005
20	0.0240	0.0007	0.000005
30	0.0483	0.0006	0.000008
40	0.1065	0.0009	0.00002
50	0.2148	0.0012	0.00003

The exponential fit of the trend can also be explained by looking into the Henderson-Hasselbach equation stated in the introduction. As the concentration of CO_2 increases, the concentration of the weak acid also increases. This will cause the pH of the seawater buffer to change due to the change in the ratio of [conjugate base] and [weak acid]. As seen in equation 5, as the ratio decreases, the overall pH will decrease, producing more H^+ ions that will react with calcium carbonate to produce even more Ca^{2+} ions ("Option D. Medicinal Chem"). This is seen in the collected results, where the concentration of calcium ions was significantly higher with a carbon dioxide concentration measured at 50 ppm compared to 10 ppm (which is the standard atmospheric concentration) with results at 0.018 M and 0.215 M respectively. The difference between independent variable also had an increasing value.

5. Evaluation

The overall experiment followed a strong and precise methodology, with minimal random error, seen with the standard deviation. However, given that the standard deviation was higher than the random error, there are factors that should be considered for a better future methodology, possibly because a control variable was not enforced.

Firstly, the biggest systematic weakness could have been the possible reaction with other chemicals in seawater, which included other ions and organic compounds. Especially since I used real seawater to best emulate real life conditions, it was not possible to control the presence of other chemicals inside of such. Possible interactions could mean that in the titration, other chemicals were factored into the final concentration. One aspect was already addressed in the experiment with the addition of anhydrous magnesium chloride hexahydrate to bind to the magnesium ions in seawater as a masking agent. In other words, I precipitated the magnesium ions to avoid them interfering with the reaction by reacting it with EDTA. For a future experiment, I could do preliminary research on other chemicals present in seawater, such as potassium and sodium, and find their respective masking agents to add in the experiment.

Even with the emulation of the seawater, the experiment did not assume a perfect simulation of seawater conditions, especially abiotic factors. Instances such as temperature, pressure, and the interactions of marine life were not taken into consideration, thus leading to another systematic error, where it could either quicken the dissociation of calcium ions or not, which would need to be tested. However, this is a relatively minor aspect, and could be easily changed by constructing a mesocosm to do the experiment in. Considerations could also be made given how each locality or geography yields various natural pH values, and thus adjust accordingly.

Possibly the largest random error was the difficulty in determining the titration endpoint as the color change from wine pink to a faint blue was very minimal and gradual. Therefore, the volume of the titrant would have a lower accuracy and possibly be larger than expected, given how the eye detected the color change slightly after it has already occurred. Though it would be difficult to fix this issue, if time permitted, the titration could be conducted slower, allowing for time to see if each drop led to a color change. Furthermore, there are other experimental methods for quantifying calcium ion concentration. For example, one can use UV vis spectroscopy to measure the absorbance of Ca-EDTA complex to determine concentration (Carpenter, 2014).

In addition, the continuous diffusion of CO_2 into the sample seawater could have been better quantified. We are not able to really control the absorption of CO_2 into the water column given the solubility of gases in water is affected by several factors including surface area, temperature, pressure. CO_2 is a slightly soluble gas. The method for CO_2 absorption is actually an approximation based on the user manual rather than a definite measurement. Though it may be said that the amount of CO_2 can diffuse into the atmosphere, it is expected to remain consistent during the titration process as the solubility of a gas is dependent on temperature and atmospheric pressure. As long as there is no drastic change in these two factors, the titration can be conducted. In future experiments, the pH itself could be a more accurate independent variable using a pH probe given various concentrations of carbon dioxide.

Another systematic error was that in this experiment, we did not conduct the standardization of EDTA concentration for the titration, or in other words, the exact concentration of the titrant solution. This affects all trials equally and wouldn't affect the trend of the results as a whole, however, it would still give an inaccurate final result by a small margin. This standardization procedure involves the correction of measurement uncertainties, the volumetric solution, handling of chemicals, measuring balance, and environmental conditions by titrating EDTA itself without a known concentration and volume of standard calcium solution. In the future, when we obtain these values, we will be able to substitute them into the $M_1V_1 = M_2V_2$ equation stated in the aforementioned, hence finding the standardized EDTA concentration.

Extension

To further develop this experiment, one can identify rates of reaction impacted by temperature levels rising from carbon dioxide induced global warming as well as concentration of H^+ ions. We can also investigate if rising temperature can have an effect in lowering CO_2 concentration in seawater and thus changing pH and its effect on Ca^{2+} concentration.This could also offer insight to the further degradation of coral reefs as a long-term projection depending on how fast this reacts of dissociation occurs.

Works Cited

"OSH Answers Fact Sheets". Canadian Centre for Occupational Health and Safety, 20 Oct. 2019,
>   www.ccohs.ca/oshanswers/chemicals/chem_profiles/carbon_dioxide.html.

Carpenter, James H. "The Determination of Calcium in Natural Waters." Johns Hopkins University, 2014, aslopubs.onlinelibrary.wiley.com/doi/pdf/10.1002/lno.1957.2.3.0271.

Dockrill, Peter. "It's Official: Atmospheric CO_2 Just Exceeded 415 Ppm For The First Time in Human History." ScienceAlert, 13 May 2019,
>   www.sciencealert.com/it-s-official-atmospheric-co2-just-exceeded-415-ppm-for-first-time-in-human-history.

"EDTA Titration." California State University, Aug. 2008,
>   www.calstatela.edu/sites/default/files/dept/chem/08fall/201-lab/calcium_experiment_xw.pdf.

"Option D. Medicinal Chemistry." IB Chemistry, by Martin Bluemel, Oxford Study Courses, 2008, pp. 740–743.

www.ingramcontent.com/pod-product-compliance
Lightning Source LLC
Chambersburg PA
CBHW061105210326
41597CB00021B/3982